# MEN IN THE SEA

# MEN
# *in the* SEA

*by*

PETER BRIGGS

SIMON & SCHUSTER • NEW YORK

PUBLISHED BY SIMON & SCHUSTER, INC.
ROCKEFELLER CENTER, 630 FIFTH AVENUE
NEW YORK, NEW YORK 10020

FIRST PRINTING

LIBRARY OF CONGRESS CATALOG CARD NUMBER: 68-18967
MANUFACTURED IN THE UNITED STATES OF AMERICA
PRINTED BY THE MURRAY PRINTING COMPANY, FORGE VILLAGE, MASS.
BOUND BY THE BOOK PRESS, BRATTLEBORO, VT.

*The author wishes to thank Scientific American, Inc., for permission to reproduce the drawing of* Galatheahauma Axeli *from "Animals of the Abyss" by Anton F. Bruun. Copyright © 1957 by Scientific American, Inc. All rights reserved.*

# Contents


| | | |
|---|---|---|
| | INTRODUCTION: *In the Beginning* | 7 |
| 1 | ROGER REVELLE: *"Dream No Little Dreams"* | 11 |
| 2 | ATHELSTAN SPILHAUS: *Prophet of the Deep* | 24 |
| 3 | GEORGE BOND: *Papa Topside* | 34 |
| 4 | ANTON BRUUN: *In Quest of Sea Serpents* | 46 |
| 5 | MAURICE EWING: *The Man Who Makes Earthquakes* | 58 |
| 6 | EDWIN A. LINK: *The Limits of Human Experience* | 69 |
| 7 | TOWNSEND CROMWELL: *Neptune's Child* | 80 |
| 8 | *Alvin* | 94 |
| 9 | COLUMBUS O'DONNELL ISELIN: *The Captain Was a Scientist* | 106 |
| 10 | TAYLOR ALDRICH PRYOR: *". . . the nation will enter the sea."* | 117 |
| | SUGGESTED READING | 124 |
| | INDEX | 125 |

# Introduction

## *In the Beginning*

THE MEN whose lives are described in this book belong to an aristocracy, that of the ocean. Their stories are adventures, both intellectual and physical. What these men have in common is excellence and a passionate interest in the wet, salty part of the world.

Although more than three-fourths of the world's surface is under water, until about one hundred years ago practically nothing was known about what goes on under the waves. Sailing on the ocean, men of curiosity and daring had found all the little islands and the great continents, but not until 1872 did anyone organize a scientific expedition to explore the sea itself.

Captain C. Wyville Thompson of the Royal British Navy set out with a group of scientists in 1872 on a two-year voyage around the world, aboard H.M.S. *Challenger.* At their first scientific stop, in the stormy Bay of Biscay, a dredge was lowered to measure the depth and to see what the ocean bottom was made of. The dredge line went down to 6,750 feet and then was brought up. The operation was powered by the new steam winches that had just been invented for fishing. When the dredge was brought back aboard, it contained a large, cold glob of mud and ooze. To celebrate this first catch, champagne was chilled in the deep-sea mud, and all hands toasted the success of the voyage.

There has been a certain amount of fun ever since in the serious pursuit of knowledge at sea. The work may be hard, the ships uncomfortable, the weather terrible, the disappointments numerous, but there is a wealth of things to be discovered, the companions are bright men who share each other's interest, great achievements are possible at almost any moment, and sailing itself is often pure joy. The sights and sounds, the smell of the sea offer something incomparable.

One of the men in this book, Columbus O'Donnell Iselin, recently wrote that "oceanography has suddenly become fashionable."

When Iselin himself began to study the subject, hardly any man on the street had even heard of the word.

What is oceanography and why has it become fashionable?

Oceanography is the use of all sciences to answer the thousands of questions raised by the sea. For example, where are the great ocean currents, and what causes them to shift their courses? How does the sea affect weather on land? What does the ocean floor look like? Why is it made of basalt while the continents are granite? How many fish live in the sea, and how many can be harvested without, like the carrier pigeon, suddenly becoming extinct? What causes the "red tide" which sometimes kills fish by the millions? Are the great glaciers of the Antarctic and Greenland that spill into the sea growing or slowly melting? Over the ages how much has sea level changed? What is the explanation for the undersea mountain range, forty thousand miles long, that loops around the globe two and a half times? Why is this range splitting in a rift down the middle? How did all the water in the ocean get there in the first place?

These are just a few of the many questions asked by men who explore the sea. There are many, many more. To find the answers to such questions physics, biology, or a combination of sciences are used. What distinguishes an oceanographer from a biologist, a geologist, or a mathematician is that he is a combination of all these things. And what is more, is probably a sailor as well.

In practice, an oceanographer usually specializes to some extent in one scientific area. His main interest may be the shape or composition of the sea bottom. He may study the immensely complex behavior of sound in water. Or he may care more about the plankton population of the sea, with a view to increasing its production. He may want to know where eels lay their eggs or if Africa and South America were once joined together. Whatever his special interest, he will understand the work of his colleagues. And he may, after some experience in the field, become interested in just about everything that has to do with the sea.

There are hardly more than a thousand people in the world today who, in academic terms, may officially calls themselves oceanographers. Many of the most prominent men in the field never studied oceanography in school at all. Two of the men in this book began their careers in aviation; another is a medical doctor. Much of everyone's learning about oceanography has come from actual experience, not from books.

Scarcely more than thirty years ago no school even awarded a degree in oceanography. Before World War II only two oceanographic institutions existed in the United States. One of them, Woods Hole in Massachusetts, had an annual budget of $150,000, not enough to run an average high school. Today Woods Hole spends millions of dollars every year. The difference is the U. S. Navy's contribution to the program.

While the Germans were stepping up submarine warfare in the Atlantic in 1940, the U.S. Navy was discovering that it knew far too little about the sounds that ships and fish made in the sea. Submarines could hide behind barriers, but a surface ship's propeller could be detected from many miles away. The Navy enlisted every qualified man it could find to study this problem. Since then, as submarines became bigger, faster, and were able to dive deeper, the Navy has increased its support for marine research. With this stimulation oceanography simply had to grow.

Just as important, though, are the new tools which have made new discoveries possible. Each advance may stimulate several others, and the whole process accelerates. When the *Challenger* in 1872 wanted to know how deep the water was, it had to let out a long line to the ocean bottom. In water two or three miles deep it would take the better part of a day to make one sounding. Now, with sonic depth finders the distance to the bottom can be found in a few seconds. Other new tools such as precision navigation equipment, airplanes which can sense ocean temperatures from the sky, computers, and underwater television have helped to speed up the pace of discovery. Men of intellectual curiosity who want to see results have now been attracted to a science which was once simply too much hard work.

There is also more equipment on hand than ever before. During the two year International Indian Ocean Expedition of 1963 and 1964, forty ships were made available for oceanographic study from a number of nations. Three of the newest ships had been built with nothing but oceanographic research in mind.

For some years there has been a great deal of talk about the sea as the answer to the world's increasing food problems and as the source for raw materials now becoming scarce on land. There has been more than just talk in the matter of oil. The oil industry is investing billions of dollars in the sea. Wells are commonly dug now in six hundred feet of water, and prospecting for new oil fields is going on in seas much deeper than this.

The ocean is being seen as the basis for better weather forecasting, for weather modification, as the source of fresh groundwater that sometimes runs beneath the ocean bottom as well as the source of fresh water through desalinization, and as a place to dump radioactive wastes. As the sea is used more and more for recreation and boating, the problems of pollution must also be considered.

Oceanographers are concerned with a vast number of things. Not all of these interests turn up in these brief biographies of nine men and one little submarine, but enough to give a good idea of the range. The sea is vast, complicated, and fascinating. The study of it is very new. And, as Albert Falco, one of Jacques Yves Cousteau's divers, said, "Under the sea, everything is moral," Exploration of it brings out a man's finest qualities. It has challenged the men in this book to a number of achievements. They are not the only fine oceanographers, of course, but their activities seem particularly interesting. And, though they have accomplished a great deal, there is a vast amount still to be learned about and dared on man's last frontier on earth.

# 1 Roger Revelle

## *"Dream No Little Dreams"*

*Men get into ocean work sometimes it seems, almost by pure accident. A specialist called in to help solve a single problem may often grow in so many directions that the entire world finally becomes his province. Such a man might be called "a scientist turned statesman."*

THE EVOLUTION of Roger Revelle from a young rock geologist into a world water-philosopher, in time and dimension, dovetails the growth of oceanography itself. Not so long ago a loose confederation of obscure sciences of interest only to specialists like Dr. Revelle, oceanography has now developed a great concern for the future of the human race. And it has come a long way from specialties toward an integration of all ocean knowledge.

While Revelle was only an undergraduate at Pomona College in California, the Scripps Institution in La Jolla formally added the word "Oceanography" to its name. At that time few people knew or cared about the change. Although Dr. Revelle finally devoted more than thirty years of his life to Scripps and its work, the new word, even if he understood it, caused no ripple in his life as a Pomona student.

Then in 1930 a yacht owned by Scripps called the *Carnegie* blew up accidentally at Noumea in the far-away eastern Pacific, and Revelle's life changed radically. A number of deep-sea cores (long, slim samples of the ocean-bottom sediments) recovered from the wreckage were returned to Scripps. The Institution needed a geologist to analyze the rock components of the cores. They called on the University of California at Berkeley, where Revelle was by then studying for his doctor's degree. He was chosen to go south to look at the samples.

Soon he found himself being sent to sea on other research expeditions. Although he remarked years later that "Man is not a seagoing animal," he never really believed it. He found these expeditions very pleasant and said to himself, "This is wonderful—you work, and take boat trips at the same time." He decided that he liked the new ocean science. Scripps offered him a job teaching geology to the students.

Even a short visit to the little town of La Jolla in Southern California makes it obvious why it became the home of the first private institution in the United States devoted to oceanography. La Jolla has a sunny, comfortable climate the year round, a beautiful beach, some magnificent views from the cliffs, and beneath the blue waters offshore, the Pacific Ocean teems with life. In addition, there is a large U.S. Navy base in nearby San Diego, which in time provided many opportunities for cooperation with the Institution.

The first scientist to locate at La Jolla, Professor W. E. Ritter, a zoologist from the University of California, did so after a number of summer field trips, camping up and down the state's long coastline. Marine animals were the object of his study. During his travels he made the acquaintance of Miss E. B. Scripps, an enthusiastic, though amateur, natural scientist. She pointed out all the advantages of La Jolla as a permanent laboratory and offered financial support from her younger half brother, E. W. Scripps, who was a wealthy newspaper publisher. Professor Ritter settled in La Jolla in 1903.

As research at the Institution developed and the importance of marine biology became more apparent, the State of California and the U.S. Government also began to provide funds. Originally known as the Marine Biological Association of San Diego, the Institution became, as originally planned, part of the University of California in 1912. The name "Scripps" was adopted in honor of the early benefactors. "Oceanography" was later added to the title because it became recognized that there was much more to the sea than just the fish who lived in it. The emphasis continued to be on the study of marine biology, but during the years Revelle taught and did research at Scripps, the Institution's work slowly broadened to cover the whole spectrum of things on, in, and under the ocean.

Many interesting discoveries were made by the scientists at Scripps, most of whom posed questions which demanded answers. Why, for instance, were there enormous cracks, hundreds of miles long, running east and west on the bottom of the North Pacific? Why are there no sediments in the Pacific more than one hundred million years old, whereas sediments many times that age exist on land? How did the small manganese nodules in the sea, billions of them lying at depths of two or three miles, get there in the first place? How can flat-topped guyots, leveled-off mountains found at a depth of perhaps half a mile below the surface, be explained? Their top surfaces must have been flattened by erosion and the guyots, therefore, must have been above sea level at one time. Did the guyots sink, or did the sea level rise that much? Why does the Cromwell, or Equatorial, Undercurrent, exist—and flow in the opposite direction of the surface current? Where are the best places to catch tuna? Why did the sardines, which used to be California's major fishing industry, suddenly almost disappear? Along with his fellow workers, Revelle, who had long since advanced from a curiosity about just rocks, pondered these and other questions.

By the time the United States entered World War II, Roger Revelle had achieved considerable fame as an ocean researcher. The Navy came to him, made him a Commander, and asked him about the various ways that sound operates in the water, undersea warfare, forecasting the surf for amphibious invasions, and survival after disaster at sea. Then in July, 1946, the Navy sent Operation Crossroads to the atoll of Bikini in the Pacific. The object was to conduct atomic tests in which bombs were exploded beneath the sea. (For the record, these tests were the inspiration for the name of a very skimpy bathing suit.) Commander Revelle was named supervisor of the oceanographic measurements.

Revelle reported that waves 60 to 90 feet high occurred at the center of the explosion. In the vicinity of his ship, 8,100 feet away, waves ran between 10 and 15 feet. He also concluded that the tests showed that no atomic bomb could cause an earthquake or a tidal wave, but at the same time he began to be very concerned about the menace of radioactivity in the sea—not only from bomb explosions but also from using the oceans as the dumping ground for radioactive wastes left over from peaceful uses of nuclear energy.

Before returning to Scripps, Revelle made a considerable effort to interest the Navy in conducting basic research relating to oceanography, particularly in geophysics, the study of the structure of the earth. He realized that if the expensive science of oceanography were to grow, it would need as much Federal assistance as it could obtain.

Back at Scripps, Dr. Revelle soon became a full professor and Associate Director. In 1950 he participated in the Institution's first long oceanographic expedition. As scientific director of the voyage, he was in charge of the *Horizon* for part of a 15,368 mile voyage. The *Horizon* was really nothing more than a surplus tug boat built during World War II and acquired from the U.S. Navy, but it was a great advance over the sailing ships with auxiliary engines that Scripps had had up to that time. Oceanography had yet to acquire the status which commands, today, ships of 4,000 tons, specially built, air-conditioned, and capable of traveling for extended periods in any sea. In spite of its lack of sophistication, however, the *Horizon* stayed at sea for several months, studying the geology and geophysics of the deeps.

Later, when Scripps acquired the *Spencer F. Baird,* almost a twin of the *Horizon,* an expedition called Shellback was organized. The two ships were to sail to the eastern tropical Pacific. Using the methods of seismology, the science developed to study earthquakes, the major object was to learn as much as possible about the ocean bottom. With two ships on hand, the most refined kind of seismic shooting would be possible: one vessel would act as the silent audience while the other, miles away, would drop depth charges. With sensitive sound gear the listening ship would time the return of the echo from the apparent ocean bottom (the surface of the sediment) and then the slower return of an echo from the real crust of the earth (the mantle beneath). The difference in the speed of the two echoes would give the depth or thickness of the sedimentary layer.

The convoy sailed on September 26, 1953, with the object, particularly, of studying the little-known Tonga Trench in the South Pacific. Dr. Revelle was scientific leader for the first two months. It had been planned that several of the scientists' wives would fly out to join their husbands in the Pacific for Christmas, but finally only Helen Raitt, wife of Russ Raitt, senior geophysicist, made the trip.

Mrs. Raitt later wrote a charming book about her experiences. In it she tells of her frantic efforts to find the *Baird* among the many Tonga Islands, hampered by the very meager communications facilities between them. While looking and waiting, she found that she had to explain her presence and her husband's work to the natives. "He is a man who makes his own earthquakes and studies them," she said. What she hoped to get across was the idea of seismic shooting.

Finally the *Baird* appeared, and since there was room for her,

RICKY GRIGG, UNIVERSITY OF CALIFORNIA AT SAN DIEGO

*Roger Revelle examining a radar reflector for the drogue float used to measure ocean currents*

Dr. Revelle suggested that Mrs. Raitt stay aboard and cruise with them rather than just see her husband briefly whenever he could get ashore.

Thus, Mrs. Raitt was present when the *Baird* and the *Horizon* made the exciting discovery of the true depth of the trench. Echo sounding produced a reading from one spot 35,640 feet below the surface—just a bit less than 7 miles straight down and only 200 feet less than the depth of the Marianas Trench. The two ships ultimately found that the Tonga Trench is about 1,000 miles long and between

15 and 30 miles wide. They did not, however, solve two oceanographic mysteries: Why the deep trenches all around the world are approximately the same depth, and why the trenches exist in the first place.

During the Shellback Expedition, the oceanographers also found a seamount four-and-one-half miles high whose sides were almost vertical. (Seamounts are believed to be underwater volcanoes not quite strong enough to have reached the surface. But a volcano able to rise more than 20,000 feet against the pressure of the sea must have had quite a bit of power behind it.)

The scientists also discovered a line of recurring deep-sea earthquakes that begin on the bottom of the Tongo Trench and extend west 500 miles to the Fiji Islands. Everyone knows that a vast amount of heat comes from beneath the earth's surface by means of volcanoes, but actually everywhere on earth a certain amount of heat comes from below. While measuring the heat given off by the floor of the trench, they seemed to have found that it was about the same as that given off by land. This would have dealt a blow to the theory that heat from land masses is caused by the radioactive elements in continental granite. The basalt from which ocean floors are composed does not contain these radioactive elements. However, these findings later seemed incorrect.

Writing about the trip to the Tonga Trench, Revelle delivered himself of a definition that has often been quoted. "Oceanographers," he said, "are essentially just sailors who use big words." He also asked some basic questions of his science. "Why is there an ocean? I wonder why water is here and how it arrived. What is the bottom of the ocean like?" This at least he could answer. "It is rough, hilly and rocky and, elsewhere, flat and muddy." He explained the basic object of the expedition—to study the "ocean floor which gives the earth's history."

Speaking as a scientific philosopher, Revelle gave three injunctions to younger men. "Ask questions of the ocean. Think no small thoughts and dream no little dreams." On another occasion he once remarked, "The only people who really accomplish something are those who work on problems they have themselves conceived."

About the South Seas, he agreed that they "are wonderfully romantic" but that "it is rather hard, grim work at sea." And he disputed Dr. Samuel Johnson, whom he quoted: "No man need go to sea who has the wit to get himself in jail." For himself, Roger Nevelle replied,

"I have often wondered why it is so pleasant to be on a small, oily, uncomfortable ship—far from the nearest land. I am convinced that it is because on shipboard the past and future disappear—only the present is left." In 1952, the year before the Shellback Expedition, Dr. Revelle had become Director of the Scripps Institution of Oceanography. Administrative duties did not allow him much time for oceanographic voyages or for basic research. The scientists in his charge, however, increased the world's knowledge on many fronts.

Following up those disturbing measurements around Tonga, which suggested that perhaps radioactivity was not the cause of heat flow from land, Dr. Richard Herzen, with a new instrument, made hundreds of new bottom measurements in the South Pacific, South Atlantic, and Indian Ocean. He found the heat flow from trenches much like that previously recorded, but the heat flow from the tops of the great ocean ridges forty times as great. This coincides with the conclusion made at the Lamont Geological Observatory in New York—that the heat and pressure from the earth's deep mantle are slowly spreading the earth apart at the center of the ridges and increasing, almost imperceptibly, the surface of the earth.

The surface of the earth has a certain amount of magnetism, just as a piece of iron does. Using a magnetometer, which reacts to the same forces as an every-day toy magnet, Professor Victor Vaquier and his colleagues at Scripps discovered that parts of the earth's crust have slipped past each other by as much as seven hundred miles. This finding, accomplished by matching the direction of magnetic force, shows that parts of the earth's crust that should line up, like the pieces of a jigsaw puzzle, have moved far apart from their original positions. The evidence that such a large motion of the earth's surface occurs shows that it is perfectly possible for continents to change their positions and that the theory of "Continental Drift"—the idea that South America and Africa, for example, were once joined together—is not at all improbable. If one examines a map he will see why the idea seems quite reasonable.

Professor H. W. Menard showed that fifty to one hundred million years ago a mid-ocean ridge once ran from off the central coast of South America nearly as far as Japan. It has disappeared now, but proof of its former existence comes from seamounts and coral atolls that once were volcanic islands on the crest of the ridge.

Townsend Cromwell, a graduate of Scripps, found an east-moving undercurrent at the Equator, which surprised the theorists. Dr. John Knauss, then at Scripps though now at Rhode Island University, investigated the phenomenon further, verified it, and found that its flow was many times the average volume of the Mississippi River.

The sudden disappearance of the sardine population off the coast of California had been bothering Scripps even before Revelle took command. Professor John Isaacs finally concluded that what happened was a simple matter of over-fishing one species. Anchovies, which compete with sardines for the same food and living space, were ignored while humans went after the sardines. As their fishy competition declined, the anchovies thrived and are now hundreds of times more abundant than the formerly dominant sardines.

While these and many other discoveries were being made, the Scripps Institution under Roger Revelle continued to grow physically. Its fleet increased to seven vessels, including a strange research ship called *Flip*. This ship can "flip" itself up and stand on its tail, so to speak, so that 300 feet of its length are underwater and 55 feet remain above sea level. It is used for underwater sound research and it can hold steady in extremely rough weather. The other research vessels are comparatively ordinary, but under Revelle's direction they made a total of forty-one long voyages and covered a total distance of 791,683 miles.

Scripps, which is part of a state university as well as a research center, grew to a campus of nine buildings on 158 acres, with a radio station, a 1,000-foot pier, an underwater research station off shore, an enormous library, several thousand samples of sea water from all over the world, one hundred thousand specimens of fifteen hundred different kinds of fish, an elaborate computer system, thousands of deep-sea cores, many finely equipped laboratories, and an aquarium and beach open to the public.

Even though faced with all the executive problems of this establishment and its nine million dollar annual budget, Dr. Revelle found it impossible to remain completely inactive as a scientist.

Early in 1953 he announced that the North Pole was moving south. Judging from measurements taken over the years, Revelle concluded that the Pole had moved ten feet between 1900 and 1925. He attributed this motion to the slow melting of the ice caps of Greenland and the Antarctic. This, in turn, raised the sea level. More water in the sea slightly alters the distribution of matter and mass on the planet, lengthens the days, and displaces the poles.

Dr. Revelle did manage to get away from his desk long enough to take part in some expeditions across the Pacific. On one of these trips the depth recorder of the *Spencer Baird* kept picking up seamounts. When samples of the tops of the seamounts, secured by drags (containers with shovel attachments) hung down from long lines, were brought up, they contained fossilized relics of clams, snails, and coral. These are species of animals which can live only in shallow water—at the most, one hundred feet below the surface. The seamount peaks were thousands of feet below the surface.

Wondering about this, Revelle proposed the theory that volcanic explosions forced the collapse of the Pacific Ocean floor about one hundred million years ago. These explosions, he thought, released volcanic spring waters that increased the ocean's volume of water by between 25 and 30 per cent. "Volcanic basalt gushed out from the earth's suffering crust and swept the Pacific clean of all previous sediment." So, this assumed explosion nicely explains not only the presence of the fossilized relics but also why the sediments in the Pacific date back only one hundred million years.

Revelle found additional evidence for his idea from his earlier investigations around the Pacific atolls of Bikini and around Eniwetok. These ancient volcanoes "seemed to have been drowned in water released at the time of the great collapse."

The same explosion theory offered Revelle an explanation for a kind of sediment frequently found on ocean bottoms. The blow-up caused primitive protozoa to become gas-tolerant and able to use the carbon dioxide just released in the water. These tiny animals lived their short lives, died, sank to the bottom, and formed the stuff now known as globigerina ooze.

To account for the geophysical changes going on constantly in the earth, Dr. Revelle proposed, in a paper he read to a meeting of scientists in Los Angeles in 1955, that the earth's interior is slowly churning over in giant blocks, hundreds or thousands of miles in size. The movement is incredibly small, perhaps one mile in 100,000 years. It is caused by radioactive heating at great depths which makes rocks far below move upward in some places and downward in others. This internal heat is of "fundamental importance in shaping the earth's surface." Volcanoes, earthquakes, mountain ranges and ridges, and deep-sea trenches—all result from slow convulsions within the earth caused by the escaping heat.

Continuing with the idea, Revelle said that research is being done to find out if a pattern exists in the quantity of heat flowing outward from different areas. Referring to work done at Scripps, he suggested that a pattern may have been found. Near the Great Acapulco Trench off Central America the heat flow is much lower than the average. Five hundred miles to the west, where there is a high ridge rising above the sea floor, the heat flow is twice that of the average. It may be that the rocks are moving upward under the ridges and downward in the trenches.

He also discussed the possibility that the surface of the earth has been cooling since its original formation, even though radioactive heat continues to be generated from the interior. He likened the earth's skin to that of an apple as it dries. In this comparison, mountain ranges and deep trenches are wrinkles produced by the shrinking of the fruit.

Beyond such theorizing, of course, Roger Revelle still had his work to do as head of a large scientific organization. In an interview one time he called himself a "traveling salesman. My job essentially is to thrill, excite, irritate, or shame people into supporting oceanography." That interviewer suggested that what "Dr. Revelle is selling might be called the *new* oceanography, which treats study of the seas as a comprehensive, integrated science in the mainstream of human affairs, rather than as an intermittent curiosity of adventurers and scholars."

Dr. Revelle was further quoted as saying, "In topography [of the ocean floor] we're about where dry-land knowledge was in 1750. Our most detailed knowledge is one or two percent of the sea floor—off the Pacific Coast—that has been systematically mapped with soundings five miles apart. In history—geology, rocks, fossils—we're about where we were in land knowledge seventy-five years ago. As to what goes on *under* the earth's crust, we know more about ocean areas than land areas—although just scratching the surface."

The interviewer described his subject as a man whose "massive face is usually solemn" but whose "manner is amiable, modest to the point of shyness, and bereft of academic pretension. He speaks softly and calmly . . ." and "his discourse is colloquial, sprinkled with mild expletives and much unaffected terminology as 'little critturs.' " The writer was of the opinion that Dr. Revelle might be such a good salesman for the underwater world because "prospective customers for his wares can't miss him. He stands six-feet-three, with a heft of some 225 pounds."

As a university head, part of Revelle's salesmanship had to do with raising money. In 1954 he accepted a grant of one million dollars for Scripps from the Rockefeller Foundation for the purpose of expanding research in ocean resources. During a ceremony marking the event, he voiced the opinion that "In some respects, marine biology has lagged behind the other biological sciences. Because of advances in other fields, notably in genetics and chemistry, many problems are ripe for solution. Life began in the sea and from the sea we may learn something about life's basic processes."

In this speech he also referred to the fact that although the sea covers more than two-thirds of the world's surface and is abundantly full of plant and animal life, only one per cent of the world's food supply is taken from the ocean—in spite of the increasing seriousness of the world's food shortages.

Dr. Revelle, the young geologist who began to study the ocean almost by accident, no longer saw it just as a laboratory where he could investigate the fascinating processes by which our world has been made. After many years' experience at an institution which studied every single aspect of the sea, he realized that the most important thing about all the knowledge gained was in its relation to *people.* Many scientists assume a superior pose in which they maintain that their work has no object but the pursuit of pure knowledge. Some of the great ones, while equally eager to find basic truths, have recognized that science also has a duty to humanity.

Over the years, Revelle moved higher and higher in the scientific establishment. He became a member of the committee on oceanography of the International Council of Scientific Unions. He served on committees of the National Research Council. He sat on the Research and Development Board of the Department of Defense. For two years he acted as Science Advisor to the Secretary of the Interior. Although practically a commuter to Washington, he still managed to find time for his wife and daughters and to serve on the La Jolla Town Council and the board of the local Musical Arts Society.

His concern about the world population explosion and the

famine he feared possible led him to write a definitive article on water in 1963 for the *Scientific American.* In this he described the pollution and supply problems of the Southwestern states. He suggested that much could be done to improve the present low efficiency of water use in the area. Half of the irrigation water is lost in transport through evaporation and seepage. Only half the water which reaches the fields is actually used by plants, due primarily to evaporation. Measures that might be taken are the development of salt-tolerant plants for over-irrigated soil, the development of new mulching methods to reduce evaporation, the building of better canal linings to reduce seepage, and the control of water-loving weeds.

For the Southwest, however, he felt that the best method of saving the land might be for the area to turn from agriculture to a mainly industrial economy.

Around the world Revelle saw even greater water problems. In the tropics, technology to improve agriculture in those lands with the most rain and the most unused soil is lacking. In dry areas, irrigation often causes water-logging and salty fields. Spreading water over large areas of ground provides a fertile culture for breeding creatures like the malaria-bearing mosquito.

Revelle found the situation in West Pakistan the greatest cause for despair. There the Indus River provides water for the largest single irrigated region on earth. But there is famine in the land. The flat plains are water-logged and silted up. Too much water is lost through seepage. The farms are much too small and the agricultural methods belong to the Middle Ages.

But even Pakistan's problems could be solved by pumping off some of the large water table, by establishing water control projects, each about one million acres in size, and by adopting modern farming methods. He saw hope, too, in the desert Tigris-Euphrates Valley. He believed its fertility could be restored if the soil were leached and drained of its salt. Millions of acres could be put back in business.

It seems striking that an oceanographer should write of such apparently remote matters. Actually this article in *Scientific American* was a very broad hint that his whole career would soon change.

A report Revelle helped write a short time later also shows the turn his thinking was beginning to take. The subject was not how molten lava bubbled beneath the crust of the earth but how minerals in the sea could be put to practical use.

Manganese, a metal used to harden steel, is found in the United States only in Montana and much has to be imported. The sea bottom, for reasons yet unexplained, has millions of tons of manganese in the form of nodules usually the size of a baking potato. The report urged that the locations of manganese nodules in the deeps be carefully mapped to find the best areas for mining them. It also discussed the offshore California phosphorite, useful as fertilizer. It offered the idea that globigerina ooze might make good cement. It mentioned the sulphides that could be taken from the Black Sea, the Cariaco Trench and the Gulf of California. (These parts of the ocean are without oxygen below about six hundred feet. They cannot support life at such depths, but they are rich in hydrogen sulphide.) The report also described how the Japanese had mined seven tons of iron ore in four years out of Tokyo Bay and showed what various ores might be taken from different kinds of beach sand. Never a man in an ivory tower, Roger Revelle had very certainly become involved in wordly affairs.

In 1964, after thirty-four years of service, Revelle retired from Scripps to go to Harvard as Director of its new Center for Population Studies. Here, he is concentrating on the uses of science to better the prospects of the world in the future. An article on pollution he wrote recently is called "Can Man Domesticate Himself?" The oceanographer feels cautiously hopeful than man can.

# 2 Athelstan Spilhaus

## *Prophet of the Deep*

*Making measurements in the sea used to be a long and tedious job. It took an aviation engineer, interested in the hot and cold winds of the upper air, to see how an instrument invented to improve flying could be used to aid fishermen—and, in time of war, to help discover the enemy's submarines.*

"THE OCEAN's bottom is at least as interesting as the moon's backside."

This line, one often quoted by spokesmen for the science of the sea, originated with Athelstan Spilhaus. He said it at a U.S. Navy symposium on oceanography in 1964. Yet his own home and center for operations was, until 1967, one thousand miles from any ocean. What makes him care about the dramas going on under the waves?

Dr. Spilhaus, who has become one of the "grand old men" of oceanography, was for many years Dean of the University of Minnesota's Institute of Technology in Minneapolis. Once interested in the conquest of the air, in recent years he has assumed an increasingly active role as advocate for the exploration and development of what is commonly called "our last frontier."

Confidante of Presidents, Cabinet members, Congressmen, the chairman of four learned societies and currently a member of twenty-nine others, Athelstan Spilhaus looks more like a prosperous rancher than a statesman for science. Physically he is on the short side, rugged and tanned, with the hint of an accent learned somewhere in the old British Empire (but definitely not at Oxford).

Among his many other achievements, his invention of a temperature gauge called a bathythermograph became vital to the Allies in anti-submarine warfare during World War II and absolutely essential for all oceanographic cruises since that time. He has also developed a "Space Clock" which retails for $250 and a syndicated newspaper science feature called "Our New Age," which runs in scores of newspapers. Not the least bit stuffy, during his frequent travels he chooses not the elegant hotels his status might suggest but those willing to accommodate his several dogs. A very frank man, when Minnesota winters would become unbearably cold, he admits to having invented "oceanographic business" so that he might fly to his island home in the Bahamas.

Born in Capetown, Union of South Africa, Spilhaus comes from distinguished parents. His father was South Africa's ambassador to a number of nations in Europe. His mother was the first woman ever elected to the South African Parliament, as well as a strong advocate of equal rights both for women and the colored population. (Dr. Spilhaus has no sympathy for the apartheid laws of the country of his birth.)

At the age of fifteen the future oceanographer began going to sea as a deck hand on a freighter that plied regularly between Capetown and Bombay in India. The seagoing life did not interfere with school, however, and at the age of twenty Spilhaus had earned his Bachelor of Science degree at the University of Johannesburg. At this age it seemed to him that the future of his country, with its vast distances, lay in aviation. With a talent for the mechanical and an interest in the future, he resolved to study aeronautics.

In 1931, the year he graduated, Germany was leading all of Europe, and perhaps the world, in aviation technology. Spilhaus went there to study and to work at the factory which manufactures Fokker aircraft. He did not like what he saw. In direct violation of the Versailles Treaty, the Germans were building war planes and, while not actually arming them, were storing machine guns and bombs alongside the planes, which could be rapidly fitted on for action. Spilhaus' sympathies were with England, one of the obvious targets for this armament.

He left Germany and came to the United States, partly, he says, "Because I'd already seen all of Europe while my father was an ambassador." In 1931 only two colleges in the United States, the Massachusetts Institute of Technology and the California Institute of Technology, were teaching aeronautical engineering. Because Boston was closer to Europe, Spilhaus chose M.I.T. When he arrived he found that the classes were all filled. He explained to the authorities that he hadn't money enough to go either to California or home. They took pity on him. He was allowed to attend classes on probation and he quickly earned a place as a regular scholar.

While studying aviation theory and technology, he became particularly interested in meteorology, the science of weather. After he obtained his master's degree in 1933, Spilhaus returned to South Africa, where he was employed for a time by the country's defense forces. During that time he also initiated the first investigations of weather in the upper air that had ever been made in that part of the world.

While still at M.I.T. he had learned of the work at the brand new Woods Hole Oceanographic Institution in Massachusetts. Always curious about scientific frontiers, Spilhaus had spent some time there and had come to know the men in charge. In 1936 the Oceanographic Institution wrote to him in Africa, offering him a job as research assistant, which he promptly accepted. The following year, at the age of twenty-six, Spilhaus made his greatest technical contribution to the science of the sea with the invention of the bathythermograph.

One of the curious properties of water in the sea is that sound moves through its density at least five times as fast as it does through the air in the upper atmosphere. This fact presents both problems and opportunities. Sounds of submarine propellers can be heard with the proper instruments from great distances away. Destroyers can thus track submarines by sound. At the same time, sounds from merchant ships on the ocean's surface can also be heard by lurking enemy submarines lying in wait to torpedo them.

The speed of sound in the water increases, however, if the temperature, the saltiness, or the density of the water increases. But sound passing through water is stopped or severely restricted if it runs into a thermocline, which is a sharp natural boundary between areas of warm and cold water. If a submarine is able to find a thermocline, it can hide in it, invisible to the sound gear of a searching ship.

When Spilhaus invented his bathythermograph, though, he was not thinking of military applications. He wanted an instrument that could make a continuous record of temperature against pressure, or depth, and help biologists, the fishing industry, and oceanographers locate thermoclines, which besides providing cover for submarines are areas where fish concentrate. He dreamed of a tool that, instead of being laboriously lowered into the water by hand or winch and then being brought up again to be read, could be towed behind a ship while the vessel steamed along, even at high speed. He wanted something small, fast, and easy to evaluate. He turned to his first science, meteorology, and adapted a mediagraph, an instrument that measures temperature against pressure in air, so that it could be used in the different atmosphere of the sea.

Spilhaus tested his bathythermograph on August 27, 1937, while aboard the *Atlantis,* a research sailing ship then owned by the Woods Hole Institution. After some adjustments, this first model of the tool worked gratifyingly well. Remarkably sudden changes in temperature were found in the thermocline. Great changes were found in it vertically, especially during the night. Oceanography had a valuable new instrument for studying, among other things, the currents in the sea. Fishermen could use it to find the kinds of water preferred by various species of fish.

Spilhaus took out a patent on his new instrument; but when England and the nations of the British Empire went to war in 1939, he gave them several hundred models for their destroyers and submarines and allowed them to put it in mass production without any royalties. When the United States joined the war, it also began to make bathythermographs, or BT's as they are usually nicknamed, but did not bother to reimburse the inventor. Since other holders of patents useful to the U.S. war effort were being paid royalties (to which they were legally entitled), Spilhaus felt he was being used unfairly by the government, and later he sued and won. Numerous modifications of the BT have been made since that time, but the basic credit remains his.

In 1938 Dr. Spilhaus left Woods Hole, although he has maintained an association with it to the present day. He joined the faculty of New York University and there started a Department of Meteorology. He remained officially its chairman until 1947. During World War II he went on leave to join the United States Air Force. In the field of technology he did research on, and development of, meteorological equipment. He also did work on the radar and radio search of upper air and on instruments that would take meteorological measurements from airplanes in flight. Craving action, he managed to be stationed for a time at Yenan in China, where he lived in a cave and radioed weather information to the U.S. forces, later teaching Chinese guerillas to do the same.

By the end of the war he had become a lieutenant colonel, and in 1946, the year he became a United States citizen, he was also awarded the Legion of Merit. Although still part of the N.Y.U. faculty, Dr. Spilhaus returned to South Africa during 1947 and 1948, reorganized the nation's meteorological services, and received a Doctor of Science degree from the University of Capetown. In 1949 he became Dean of the Institute of Technology at Minnesota, a grouping of faculties that includes engineering, physics, and architecture.

The administrative job did not prevent his career from moving ahead on many fronts. In 1951 he was scientific director in charge of studying the effects of atomic weapons during two tests in the deserts of Nevada. The following year he went to Korea as a member of a civilian scientific mission sponsored by the Army. He was also a United States representative at a Pacific Science Congress in New Zealand and a U.S. representative on the Executive Board of UNESCO in Paris.

During the International Geophysical Year in 1954 he made a journey to the Antarctic to do meteorological research. To get better views of an eclipse of the sun he later went to a remote spot called Kidal in the Sahara Desert. The following year he observed another eclipse from an aircraft while flying over the state of Maine. President Kennedy named him U.S. Commissioner of Science during the Seattle World's Fair. He also serves on the board of directors of four corporations and has been consultant to General Electric, General Mills, and the Minneapolis-Honeywell Company. In 1967 Dr. Spilhaus resigned from the University of Minnesota to become President of the Franklin Institute, a science education and research facility in Philadelphia.

Dr. Spilhaus holds many patents and for a hobby devises scientific toys for children. One of the inventions he is most proud of is the Spilhaus Space Clock, "the most informative clock in the world," according to the Edmund Scientific Company of Barrington, New Jersey, who manufactures it.

Designed for the space age, the clock tells time in the major cities of the world, sidereal time (the period required for a star to make a complete rotation), the sun's position and time of sunset, the time of moonrise and moonset, the moon's position and phase, the times of high and low tides at any coast, the current stage of the tide,

FRANKLIN INSTITUTE

*Athelstan Spilhaus, oceanographic statesman and inventor of the bathythermograph* the current position of the stars over any location on earth, and the day and month of the year. It also tells plain ordinary time.

With a necessary adjustment (because the positions of the constellations were worked out two thousand years ago by astrologers on whose calculations current astrological predictions are based, and have moved since then), the Space Clock may also be used by those interested in astrology. Astronomical clocks have been built since the year 1090 when a Chinese inventor, Su Sung, built one thirty feet high. Dr. Spilhaus' clock is only fifteen inches high and can be used "in any classroom, library, club, or even at home," where it would certainly provoke conversation.

His illustrated newspaper feature, "Our New Age," which simplifies science, runs every Sunday in ninety-three cities. Criticized by some staid professors of science for popularizing material they apparently want to keep obscure and difficult, Dr. Spilhaus replies, "which one of you has a class of twenty-five million people?"

He adds that "Science is an intellectual entertainment. But the entertaining part has been largely overlooked." The column, further, "gives me an excuse for poking into things my colleagues are doing in other fields." When he went to Tokyo, the Philippines, and other parts of the Pacific during his term as U.S. Commissioner of Science, he discovered that he was known, not for all his honors and titles, but as the creator of "Our New Age."

Dr. Spilhaus has also written a number of technical papers for scientific journals that no one outside the particular area of specialization can possibly read and understand. His most serious titles are *Workbook on Meteorology* and *Meteorological Instruments and Weathercraft*. More popular are *Satellite of the Sun* and *Turn to the Sea*. He is very proud of the latter because "It is the only book on oceanography that has been translated into Braille."

A prodigy with so many interests and achievements, Dean Spilhaus always seems to be where the scientific action is. Yet, whether he is concerned with atomic weapons, investigations of the upper atmosphere, eclipses, the introduction of meteorological ballooning to Minnesota, or the introduction of advanced weather science to South Africa, he is also more and more an ambassador of oceanography to the Government, to industry, and to the public.

His enthusiasm for the sea has led him to make a very specific proposal on how the United States should go about mastering it.

Recalling that universities like Minnesota were land-grant colleges—financed by grants of land given them by the Federal Government—Dr. Spilhaus proposes that sea-grant colleges be established, whose income would come from the sale of rights, chiefly mineral, to utilize the sea bottoms that had been given the various schools.

The sea-grant colleges would be the centers for a great number of oceanographic activities. Speaking about his project at Newport, Rhode Island, Dr. Spilhaus said, "Today we are inventing the future, not merely predicting it." He pointed out that the nation is running out of valuable coastline and "We should scoop it out and lengthen it."

Coastal waters should be heated up so that more of them could be used for swimming, Dr. Spilhaus believes. Useful fishes would be transplanted to these warmer waters to breed and multiply. The waters could be warmed by the waste heat from nuclear reactors. The same energy could be used to dredge sand for beaches. Many beaches are now becoming increasingly narrow because soil erosion from land is under good control today. Eroded soil is the main source of sand.

Dr. Spilhaus has visions of a great many other things that could be developed through work at the new colleges: sea airplanes for instance, that can land in the sea, do a job under the water or on the surface, then take off perhaps vertically, and go on to do a job somewhere else. He also sees cities floating under the water, perhaps 100 feet below the waves, with storm-free, artificial harbors.

Another of these college projects would be the study of soil mechanics so that sea-bottom housing could be built on tricky, uncertain ground that would be unsafe without more knowledge than is presently available. There is also need for "an expedition in crawling vehicles, containing men, across the bottom of the Atlantic and the Pacific to survey the terrain, a Lewis and Clark transocean bottom expedition." Much more has to be learned about what it is really like under the ocean. Evidence from submarine cables broken by deep water landslides shows that "there are catastrophic phenomena much stronger than our concept of 'quiet in the deeps' would lead us to believe." The end of this new knowledge is that "Once people can live and work at 1,000 feet down, the whole Continental Shelf (the sea bottom, consisting of continental granite, out to the 100-fathom curve) is opened up to us as a new continent for our use."

Dr. Spilhaus also has in mind studies on the use of differing water temperatures—the thermal gradient—as a source of power from the sea. Differences of as much as ten degrees of temperature occur in small areas. Since hot water displaces cold water, it exerts a force, a force that might be used to move electrical turbines. The technology for doing this has not yet been worked out, but it is entirely possible.

Dean Spilhaus has a number of bold ideas about sea farming, too. Living things in the sea would be made to work for us, just as we use horses and oxen today. He even believes that whales could somehow be taught to herd shrimp. The activities of coral that build reefs could be accelerated to create more land. We could also plant seaweed to stabilize beaches that are now being eroded by strong currents and storms. We might create hybrids of plants now growing in the sea to make them more useful or palatable to man. We could create ocean currents to stir up minerals on the ocean bottom and thus provide richer food for the fish population. Such a current as the Humboldt, off western South America, even today one of the world's greatest fishing grounds, might by the use of nuclear energy be induced to carry even more minerals from the floor of the sea to the surface. We could also intervene in the sea food chain. The sea food chain is the unfailing law of the sea. Small fish eat tiny animals and sea vegetation, then are eaten themselves by larger fish which, in turn, are the prey of even larger fish. We could eliminate those species not useful to mankind, the same way a farmer takes out plants he does not want—a process Spilhaus calls "Weeding the Sea."

Oceanographic scientists working with the fishing industry would become the equivalent of agricultural college county agents "in hip boots."

"This is what ocean engineering can do. We need to fill the gap and bring the vast body of knowledge of the sea to the people who use it. The sea-grant colleges would focus all intellectual activity on the ocean. They would also serve as information and communications centers to collect and index the sum total of international oceanographic knowledge."

The proposed method of supporting the colleges financially would also solve one of the most difficult and embarrassing problems about the undoubted mineral wealth at the bottom of the sea: the question of who owns the manganese, the phosphate, the oil that is down there. How does a mining company stake out a claim? Until corporations are certain of their legal rights, they are going to be very cautious about spending the millions of dollars necessary to bring the minerals to the surface. If they could lease a particular area from a sea-grant college, however, with a sure claim to the products they produced, deep-sea mining might move rapidly ahead.

Athelstan Spilhaus has been very actively pressing Congress to do something about his novel idea. He has the support of a number of Senators and Representatives, but because there is such a considerable amount of money involved and because the legal area is quite murky at present, the passage of such legislation is difficult.*

The Russians have already publicly announced that "whoever masters the sea masters the world." In his speech at Newport, Rhode Island (whose state university was making a strong bid to be the first such college), Spilhaus also warned, "We will master the sea only when we occupy it."

A man of such conviction is likely to be hard to silence. In addition he has an asset in urging Federal support of the proposed colleges that is somewhat unsuspected. Since he was a dean of a university at least one thousand miles from the nearest ocean when he began his drive, he is above suspicion. No one can suspect him of special pleading. Minnesota would gain nothing from sea-grant colleges. The nation would, however, and so would oceanography.

Athelstan Spilhaus is the ideal and eloquent partisan. He has said, "The oceans offer us military, recreational, educational, economic, artistic, and intellectual outlets of unlimited space. Thus, they offer us more than [outer] space itself in which to remain human. The sea . . . beautiful, elegant, strong, dangerous and whimsical, challenges us. . . ."

* *Although a token National Sea Grant College and Program Act was passed by Congress in 1966, its limited version is to provide a total of one million dollars a year for existing institutions to direct their programs toward technology. Dr. Spilhaus' goals are much greater than this. The bill, in particular, does not provide support to the schools through leases of the rich bottom of the sea.*

# 3 George Bond

## *Papa Topside*

*Physicians are traditionally supposed to heal the sick, but if a doctor with an active mind is sent down into the sea with a crew of particularly healthy young sailors, it is likely that he will take the free time to think. One such man's thoughts and research have produced such changes in man's approach to the sea that, although very well documented, they still seem incredible.*

GEORGE F. BOND, a Doctor of Medicine who is also a captain in the United States Navy, might well wear a miniature aqualung as his insignia of rank. Now fifty-four, proudly the oldest qualified diver in the Navy, Captain Bond is the man whose physiological research in breathing mixtures of helium and oxygen led to the successful underwater-living experiments of Commander Jacques Yves Cousteau in his Conshelfs and the various Sealab programs of the U.S. Navy.

It was through the application of Bond's studies that the astronaut Scott Carpenter was able, in the summer of 1965, to live and work at a depth of 210 feet for thirty days off La Jolla, California. Bond himself holds the world record for a free ascent from the deep, a record made in 1959 when he and a companion left the U.S. submarine *Archerfish* through the escape hatch, at a depth of 322 feet, outside Key West, without any equipment at all except life jackets, and rose safely to the surface. Hardly a stunt, the experiment showed from what depths a sailor could safely escape from a stricken submarine.

As a matter of some curiosity, Bond got into all this watery work completely by accident. He originally planned to live as a country doctor. He received his medical degree from McGill University in Canada, and as had been his ambition, with several other doctors he started a small private hospital in Bat Cave, North Carolina.

At the end of World War II he held a U.S. Army Reserve commission, which he neglected to resign. At the time of the Korean conflict, he found to his surprise that he had orders to report to Fort Sam Houston in Texas for active duty with the paratroops. But, as he says, "Somebody pushed the wrong button," and he learned that the paratroopers had no need for another doctor. The military, however, were reluctant to let a live doctor go once they had their hands on him, and they gave Dr. Bond a choice of other services he might join. The only other service offering extra pay for hazardous duty was submarines, and since he had a growing family to support, he asked to be sent to the submarine base of the Navy at New London, Connecticut.

After taking his first dive, Bond says, "I was hooked." Except for one more brief fling at civilian life, he has been in the Navy and concerned with life under the water ever since.

A big, robust man who enjoys physical challenges, Dr. Bond has a mind of the same cast. Not satisfied merely with the demands of his unusual medical practice among men who sometimes live submerged for as long as two or three months, he began to speculate on the further implications of such an existence. As a result of his experience in subs and as a scuba diver, he started to wonder about living in the sea itself, outside a submarine with its simulated sea-level atmosphere. He knew about the short effective working time, about five minutes for every hour spent in the water, that a "hard hat" or a scuba diver has when down deep. He also knew about the dangers of nitrogen narcosis, since he himself has had the bends (or "been bent") four times. He reasoned that if a diver could avoid the necessity of staying in time-consuming decompression chambers at the surface, he could do a great deal more work, more economically and much more safely.

The result of his thinking was a formal proposal to the Navy in 1957 to do basic research on the subject. The response was indignant. What right had a mere doctor to make engineering suggestions for work on the sea bottom? He was told firmly to desist.

Dr. Bond is grateful now for the refusal. He feels that the best research is often done under conditions of hardship and that opposition forces a researcher to refine and perfect his work so that finally there can be no argument about its validity.

Thus, without official sanction, but aided by Robert Workman

U.S. NAVY

*Captain George F. Bond (with pipe) with Commander Scott Carpenter and other associates in the Sealab II experiments*

and Walter Mazzone, two Navy doctors who shared his enthusiasm, he conducted experiments for five years on "Prolonged Exposure of Animals to Pressurized Normal and Synthetic Atmospheres." First they subjected a group of young male rats to pressurized air, which has the same composition as the air we breathe, at a simulated depth of 200 feet of sea water. As Bond expected, under conditions of pressurized air, all the rats were dead at the end of thirty-five hours. Next, a group of rats and guinea pigs were exposed to an atmosphere that was 97% nitrogen and 3% oxygen, at the same simulated depth, for fourteen days. Almost all the animals lost weight, the rats became sluggish and dragged their hind legs (although the guinea pigs remained active and alert), but only one animal died.

In the next experiment a group of rats breathed an atmosphere that was 80% helium and 20% oxygen. All survived well. Later successful experiments with squirrel monkeys and with goats proved that higher mammals could survive in the deep while breathing such exotic mixtures. Two problems were solved. Nitrogen, which causes the painful condition called nitrogen narcosis, was eliminated, and a usable substitute found in helium, which is non-toxic, non-explosive, and non-inflammable. Dr. Bond comments that "No other research I have ever tried has been so successful. I have to attribute most of it to just plain luck."

Even though the Navy took no interest in this work at the time, news of it reached Cousteau, inventor of the Aqualung, in France. Cousteau understood the advantages of Bond's concept of living deep in the sea. He knew how tedious it was and expensive in time to have to return to the surface and undergo decompression after only brief intervals down in the sea. He asked for, and Bond freely gave him, research figures and the data on the kind of air needed for underwater living.

Cousteau had a chamber built in 1962 which he called Conshelf One. It was carried out to a position off the coast of Marseilles and lowered forty feet into the sea. Two divers swam down to it and lived there for eight days. They swam out through an open hatch in the bottom (through which water could not enter because of the pressure inside) in their Aqualungs and often worked their way to a depth of 80 feet.

No physical ill effects were felt from living in Conshelf One, but the divers were under constant surveillance by TV from the surface, and they showed signs of irritation about being spied on and from the too frequent arrival of visitors from above. The only physical problem the men really had was in talking. Their speech resembled that of Donald Duck. All the vowel sounds disappeared. The change was caused by the fact that the speed of sound in helium is five times as great as it is in normal air. The same effect can be produced by playing a record on a turntable that is whirling too fast.

The following summer Cousteau had an entire village built underwater in the Red Sea at the same 40-foot depth, but with another chamber 80 feet below the surface, where two men lived for a week. From this lower housing they dove to a depth of 330 feet, returning to their chamber without danger of narcosis. Conshelf Two also boasted an underwater garage for the *Diving Saucer,* a miniature submarine in which Cousteau and a companion dove down to 1,000 feet. The serious problem encountered in the deeper housing came from the miserably high humidity. It was so damp that the men were more comfortable in the water than inside their shelter.

Meanwhile, waiting for Naval recognition of his work with exotic atmospheres, Dr. Bond conducted several experiments that did have official sanction. One of these was to show how pilots might escape from the cockpits of jet aircraft that were sinking in the sea after an accident. To do this he had simulated aircraft assemblies built on the afterdecks of two submarines. Bond himself was the test subject.

Wearing full jet-pilot gear, the large doctor entered the cockpit section of an F4D that had been secured to the submarine U.S.S. *Toro,* which then submerged to a depth of 70 feet, running at a speed of nine knots. There Bond gave a bail-out signal, ditched his aquamatic mouthpiece, inflated his life vest, and escaped from the cockpit.

He ascended to the surface using the "buoyant ascent technique" in which a man blows out most of the air in his lungs initially, then continues to blow air out as the life jacket carries him swiftly to the surface. A number of such tests were done in Long Island Sound. Then Bond did eighteen more tests from the deck of the submarine U.S.S. *Seacat* under the water near Key West. Only one accident occurred and it was a near fatality. Bond momentarily lost his life jacket but recovered it in time. He successfully proved that a pilot could escape even if his airplane had sunk to a depth of 100 feet.

Using the same buoyant ascent technique, Bond, along with three other Navy men, also demonstrated how men might escape from a sinking submarine. Emerging from the submarine, all the men came unharmed to the surface from depths of 100, then 200 feet. Finally, Bond set his record, with Chief Engineman Tuckfield, of 322 feet below sea level. Bond believes he could do the same thing safely from 600 feet, but so far he has not tried it.

It took the highly publicized loss of the submarine *Thresher* in 1963 in 8,000 feet of water for the Navy to become actively interested in deep-sea rescue work. With the thought that men could move out to help a sunken submarine from undersea housing, Dr. Bond's proposals now began to be taken seriously. It no longer seemed outrageous that a doctor should have ideas about living and working deep in the sea.

At New London in 1963 Dr. Bond was authorized to conduct a test in which three human beings lived in a pressure tank, breathing an atmosphere which was primarily helium for twelve days at a simulated depth of 200 feet. The success of this program led to the project known as Sealab I, and Dr. Bond, now a full Captain, in the following year was finally about to carry out his program for a real-life "Man In The Sea."

The site chosen was Argus Island, not an island really, but a tower built in 192 feet of water on top of a seamount, an extinct volcano, near Bermuda. The original purpose of the tower was military, to monitor underwater sounds made by submarines, but Bond used it as his surface control station. Sealab I, a cigar-shaped, forty foot tube, was to be lowered to the ocean bottom beneath the tower.

The four Navy volunteers chosen to live in Sealab had already trained for such an existence in special simulation chambers. While practicing, they became bored with the carefully balanced but dull meals prepared for them. They asked Bond, whom they had nicknamed "Papa Topside," if they could have more interesting food to eat during the actual test. Papa Topside did not consider what they ate of particular importance so he gave one of the men cash to buy food at a Bermuda supermarket. This man had an appetite for Mexican cooking, and he spent the entire allowance on cans of enchiladas, tacos, tortillas, and tamales. He failed to mention the fact to anyone.

While Sealab was being lowered for the first time, it accidentally flooded. By the time it was flushed out, all the labels had floated off the cans. Not only did the four divers have to live off Mexican food for breakfast, lunch, and dinner during the time they spent below, but they were never sure which kind of Mexican food would appear until they opened the can.

In Sealab I the problem of talking in helium was partially solved by the use of a special filtering device, a speech modifier, just developed by the Navy. The device improved underwater-to-surface communication, but the men below still had difficulty talking to each other. Humidity, which had troubled Cousteau's men, continued to be a problem as well. A somewhat unexpected difficulty arose from the coldness of the water when the men were outside, swimming about. They wore "wet suits" of black rubber, supposed to insulate them. But helium, the gas they breathed, radiates heat from the body five times more rapidly than air, and it was helium which filled the tiny voids in the rubber foam. The men were chilled to an uncomfortable degree. Living below, they also had headaches, sore joints, a certain amount of nausea, a slowdown in activity, and some personality changes.

"They talk," it was stated in the official report on the experiment, "often without realizing it, as though they were masters of all they survey below and above." The divers were not masters of the weather, however, and Sealab I had to be called off after ten days because of a threatened hurricane. Decompression took fifty hours. In spite of the various discomforts the men experienced, no permanent physiological ill effects have yet been discovered. Captain Bond's project had proved to the Navy that men could, ultimately, live at the bottom for months at a time, performing various naval operations, engaging in mining and fishing, farming fish, and harvesting crops from the sea.

To avoid hurricanes, the Sealab II experiment was held in the Pacific, near La Jolla and the Scripps Institution, even though the water was cloudier and colder. This housing was put down to the ocean bottom, a depth of 210 feet, while Papa Topside ran the show from the support ship, *Berkone*. Twenty-eight students from Scripps kept a twenty-four-hour watch on the men below from four TV monitors. Scott Carpenter, the astronaut, became commander below because of his enthusiasm for the project and because he had already had so much training for living in such unusual conditions. The scientists and the Navy men, including Chief Tuckfield who made the record surfacing with Bond, came up after fifteen days. But Carpenter stayed down at the bottom for the entire thirty days.

At first the divers found their high-pitched voices very funny. One of them, Earl Murray, reported: "The first day we laughed so hard we were sore." This speech problem was never more than partially solved. Neither were the numbing cold water nor the frequent ear infections. Carpenter and a torpedoman were stung painfully by scorpion fish, but a doctor on hand managed to relieve their agony. There is a story to the effect that cuts and injuries heal very rapidly in a pressurized helium atmosphere, but Dr. Bond says this is not true. It may be true, however, that beards do grow much more slowly.

Sharks were anticipated but caused no trouble.

To supplement their food, Captain Bond reported. "The aquanauts made up almost half their dietary requirements by living off the sea, eating raw fish and an amazingly tasty soup made of euphosids [a species of plankton], which occurred in such dense swarms that one could go out and with a single sweep of the plankton net get pounds of them. They tasted," said the Captain, "exactly like Wheaties."

Cousteau's divers had had strict orders against fishing because Cousteau wanted to preserve the area in its natural state, but Bond believed that Sealab disturbed the ecology so much by its presence that such an attempt was useless.

After Sealab I, Captain Bond wrote that "To observers, it appeared that the aquanauts, having become completely identified with their undersea environment, felt little requirement for surface support and began to exhibit a considerable degree of independence of action." He warned that "this 'underwater break-away' requires more complete analysis, considering the hazards involved. For some time to come, the undersea world will remain in environment hostile to man and this fact must be constantly recognized by the man who elects to live in the sea."

This undersea break-away was also a phenomenon among the men in Sealab II. Although praising Dr. Bond as "one of the world's foremost living authorities on the subject of man living and working under water," Scott Carpenter said, "He and others on the *Berkone* were in constant communication with us, and they could watch us in the lab over closed circuit television. Generally their counsel and their prompt and cheerful guidance—even their normal or earth-man voices—were extremely welcome and necessary. Yet, though we were only 200 feet apart, there were times when there was very little rapport between us."

To illustrate the problem more vividly, Bond tells the story of one aquanaut in the lab who used the electric writer, one of the means of comunication with the surface, to send a message containing his sentiments. "Good morning, Earth People," he wrote. "To hell with you."

The cold water was a continual difficulty. If the rubber suit did not fit perfectly, chilling drops of water would leak in. Sometimes, after only ten minutes, the men would start shivering so uncontrollably they would have to return to the lab and get under the hot shower. Electrically heated suits were tried out during the test, but although the idea was excellent, the suits did not work very well.

Inside the lab, it was unnecessary to put up a "No Smoking" sign. There was so little oxygen in the air that a cigarette could not burn. In the helium atmosphere, water heated even to 360 degrees still would not make bubbles. While fire is the greatest danger of all in space exploration, its impossibility below in helium is the only safe thing there is about underwater life. Another curious property of helium is that it will leak right through glass, and in Sealab II it destroyed the tubes on ten TV sets.

Among other problems, the housing had not settled on a flat bottom, and throughout the experiment, the inhabitants had to compensate for a considerable tilt. As in the Sealab I experiment, the men suffered from headaches and from the high humidity. As time progressed, their physical activity and mental functions slowed down, and a certain amount of reluctance about leaving the lab and going out into the sea gradually developed.

On the other hand, the food was satisfactory, wives and sweethearts sent down cakes and cookies, there was daily mail delivery, the men could make collect phone calls to friends and families, and on one occasion they talked with the men 440 feet under the sea off France in Cousteau's Conshelf III, an experiment that was going on at the same time.

When the men completed their long stay under the sea, they came to the surface in a transfer capsule whose atmosphere had been made to resemble the pressure at 210 feet. On deck the capsule was put in the decompression chamber and the men emerged from the capsule into the chamber to stay there for thirty-five hours while they artificially returned to the air conditions at sea level. They read, slept, or just lay around while the pressure of the air they breathed was gradually lowered and the gases in their systems, poisonous to them in the world just outside, were slowly expelled. When they finally emerged from this last ordeal, Captain Bond said, "The first thing they enjoyed was breathing the fresh air and the smokers wanted to smoke." Only one man, the oldest in the party, suffered from nitrogen poisoning, a very painful affliction of the bones and joints. Dr. Bond compares it to an "elephant's toothache." The man recovered after an extra twelve hours in the chamber. Every six months since then all the aquanauts have been given physical examinations, with particular attention to X rays of the long bones to make sure that "diver's osteitis" has not been caused by concentrations of nitrogen.

When the aquanauts were safely decompressed, medals were awarded and the Secretary of the Navy sent congratulations. In spite of the various difficulties, Bond considered the experiment a success and much better than Sealab I. "The superior performance of the aquanauts in Sealab II can be explained by their superior conditioning, higher morale stemming from the presence of a larger team, and changes in the capsule's atmosphere."

On September 1, 1965, Dr. Bond, who is at his desk in the Navy Department only one working day out of twenty, was at Ballast Point, California, for the commissioning of Sealab III. Eleven of the most decorated sailors in the Navy, all aquanauts, were there as her skeleton crew. Encouraged by the major achievement of the previous test, which demonstrated that a team of men "could live and work in the hostile environment for a prolonged period," the Navy is making plans to conduct underwater tests with Sealab III in the near future.

Sealab experiments are becoming increasingly expensive. The first one, off Bermuda, cost $148,000. The second came to one and one-half million dollars, and Sealab III cost five million. (Not at all comparable, of course, to the cost of space exploration.) The justification for this expenditure on what was until now "the vast unreachable frontier," according to Dr. Bond, is that "it is widely believed by scientists that the eventual exploration of the ocean's bottom will reveal among other things a vast new store of edible foodstuffs, petroleum and minerals at least equal to those on land, new families of antibiotics, and considerable sources of fresh water. Before all this can be done," he added, "we must know much more about the way in which men can work effectively under water and about the equipment they will need."

Hardly any equipment in a Sealab works as it would be expected to at the surface. The air pressure, the curious properties of helium, the high humidity—all do strange things to electrical circuits and machinery, to say nothing of what happens to TV tubes and the human voice. The conditions outside the lab also require considerable study. Ocean bottoms have to be selected that will not slump and must be away from areas subject to underwater landslides (turbidity currents) that could bury the housing in muck. There are also strong chemical reactions between salt water and certain metals, as well as biological effects on equipment from sea animals and bacteria whose growth is not hampered by pressure, cold, or lack of oxygen. Much more has to be learned.

One major difference between Sealab III and the earlier labs is the greater privacy given the men below. Just as Cousteau discovered with Conshelf, Bond became convinced that his men would perform much more happily without constant surveillance from above.

Asked about how he recruits his aquanauts, Bond said that the major criteria, other than good health, are that the successful aquanaut has to be interested in the science of oceanography, even if untrained in it, and understand the object of the tests. Scott Carpenter adds that "It is a profession calling for more guts and motivation than any other I know."

A number of women have volunteered, but so far Bond has resisted their appeals because appropriate facilities haven't been developed yet. One woman persisted, however, even to the extent of writing her Congressman. Bond thought he could discourage her by asking for her body measurements and photographs in a natural state. To his secretary's embarrassment, the lady promptly sent these along. Bond believes that eventually when men begin to stay down for months at a time, women will be accepted.

In future Sealab trials the ocean itself will be put to work. The water will be used somehow to scrub the carbon dioxide out of the chamber, as well as for dehumidification and refrigeration. Future aquanauts will be trained to use the ocean's sonar system. At present, sounds in the sea merely mystify divers, but Bond believes that men can, with training, become echo locaters, like porpoises, so they can orient themselves in the dark.

The doctor points out a number of similarities between space travel and deep-sea life. In both, men become disoriented. What is called weightlessness in space is called neutral buoyancy under the waves, but the sensation is the same. Sound cannot be localized, and the sense of touch is destroyed. Because of these conditions and because men cannot swim against even the slightest current at such depths, there will be less and less swimming in the future. Locomotion will be provided by vehicles.

Captain Bond believes that eventually installations like the Sealabs will be placed permanently in locations like the Hudson Canyon, off the coast of New York, along which foreign submarines can and have approached the United States. (The Hudson Canyon is one of many canyons along the coasts that is thought to have been dug by rivers when the sea level was much lower. The Hudson Canyon cuts through the Continental Shelf, whose average depth is at most 600 feet, with a gouge that goes down to 6,500 feet.) Bond foresees that some sort of housing will be built into the wall of the Canyon as a listening post which cannot be detected by a submarine but which will monitor everything that passes.

He also feels strongly that the United States should place Sealabs on all seamounts near its coasts. The Cobb Seamount, for example, not far from Seattle, rises to within 118 feet of the surface. Under present international law it could be legally occupied by any foreign power which might or might not be friendly.

Asked why the work which men can do from Sealabs cannot be done as well by miniature subs, such as *Alvin* or *Aluminaut,* Bond replied that "Man is still the best computer in existence and no mechanical hand can ever rival a human one." He believes that in four or five years Sealabs will be installed as far down as 1,700 feet and that men will routinely dive to 3,000 feet. (Other researchers doubt this and believe that the limit is about 1,000 feet.)

In the distant future Bond visualizes dives at many, many thousands of feet using "liquid breathing." This would require that the diver undergo a tracheotomy, an operation that opens the throat, and astonishingly he has twelve volunteers who are ready to try it whenever he is ready. Papa Topside inspires tremendous confidence among his men.

Asked if his work had about reached the stage that aviation reached in the 1920's, Captain Bond replied, "We're still in the age of the Wright Brothers."

# 4 Anton Bruun

## *In Quest of Sea Serpents*

*Many terrible tales have been told of monsters in the sea. The Loch Ness monster in Scotland has been good newspaper copy for many years. No movie about diving or submarines would be complete unless the hero grappled with a few murderous animals of the deep. The son of a Danish farmer sought all his life to see one of these fabulous creatures, but his hobby did not interrupt his dream of putting ocean scientists to work harmoniously together, on an international scale.*

FOR MANY YEARS, the Presidents of the United States had for their personal use a yacht called the *Williamsburg*. Always freshly painted, its polished brass gleamed when the sun shone on it. If Washington, D.C. became unbearably hot, the President and his friends could cruise on the cool waters of Chesapeake Bay. President Eisenhower apparently did not care for the sea or perhaps felt that a yacht was ostentatious, and so the Presidential flag was lowered.

When the government's National Science Foundation was looking for ships to do oceanographic research in the torrid Indian Ocean, it asked for the Presidential yacht no longer in use and, receiving it, renamed it the *Anton Bruun*. Who was the man honored in this way? Hardly anyone but an oceanographer could tell you. He was a Danish ichthyologist, a scientist mainly concerned with fish.

As the first son of a prosperous farmer, by normal expectations young Anton Bruun would one day have taken charge of his father's lands. Instead, at the age of five he fell sick with polio, and the left side of his body became partially but permanently paralyzed. Considering that his son could never lead the active outdoor life of a farmer, Anton's father sold his estate and bought another near Sero, one of Denmark's finest schools, so the boy could be near an institution where he might learn to live by his intelligence.

When first exposed to natural history, Anton became intensely interested in collecting butterflies, beetles, fossils, tropical birds, and fishes. And the boy who was not considered strong enough to run a farm taught himself to ride a bicycle. After entering the University of Copenhagen, Bruun decided on a career in zoological research, with particular emphasis on all the varied forms of animal life in the sea.

Active oceanography might seem a difficult life for a young man so handicapped, but Bruun refused to let his paralysis hinder him. At the age of twenty-two he went on a research cruise, in the Baltic Ocean and North Sea, aboard the sailing ship *Dana*—the first of many such cruises he would make with the vessel, often in rough and stormy water. He learned how to care for himself in the worst kind of weather, astonishing his friends by somehow getting in and out of small boats even while they were underway. After graduation Bruun went to work for the Danish Government's Commission on Fisheries and Marine Research. Two years later he was on board the *Dana* for a round-the-world voyage that kept him away from his family, always eager to shield and protect him, for two entire years.

The leader of the voyage was Johannes Schmidt, a scientist whose fame was based on his discovery of where eels breed. The various species of eels found in the fresh waters of Europe and North America had long been curiosities. Adults were easy to find, but no one had ever seen a young eel or observed any stage in the breeding process. Schmidt found out that reversing the behavior of salmon, eels go to the ocean to deposit their eggs and specifically to that part of the Atlantic known as the Sargasso Sea. Eels do not swim to fresh water until they are mature.

Back in 1929, when the *Dana* with Anton Bruun aboard started on its long voyage to explore the deep seas, oceanography could not be studied in schools. It could only be learned by practice, by hard work at sea. Schmidt seems to have had little interest in teaching, but Bruun was quick and observant and noticed a great deal about the techniques Schmidt had developed. He also learned much concerning the fish that swim in the sea. Blessed with an exceptional memory, he could recall, many years later, every single one of the thousands of species brought aboard the *Dana* during that entire trip.

(Interestingly, the expedition was privately financed by the Carlsberg Foundation, a scientific organization whose funds came from Denmark's largest brewery. When the oceanographers found a mountain range deep in the Indian Ocean, with peaks 10,000 and 11,000 feet up from the sea floor but 2,000 to 3,000 feet beneath the waves, running all the way from the Arabian Sea to the southern tip of Africa, they named it the Carlsberg Ridge.)

Upon returning to Copenhagen, Bruun found a wife, wrote a book about Professor Schmidt, and published the first work that brought him international attention. In this basic study, *Flying Fishes of the Atlantic,* Bruun described eighty-three different species. The monograph, which earned him his Doctor of Philosophy degree, remains today the basic work on the subject.

During the years before World War II, Bruun became Curator of the Zoological Museum of Copenhagen. He also wrote a number of papers analyzing the tremendous amount of data gathered during the *Dana*'s great cruise and a treatise entitled *Life Histories of Deep Sea Eels.* Later, a number of short cruises aboard the *Dana,* in European waters, helped increase his already enormous store of knowledge.

Ashore, he cultivated a wonderful garden at his home in a suburb of Copenhagen. Whenever he went to sea and stopped at strange ports, he would acquire whatever exotic plants he could find and nurse them in his cabin until he returned to place them in Danish soil. In time he had an amazing wealth of strange shrubs and trees to delight his visitors. And during his travels he learned to speak English, French, and German almost as well as his native tongue. Just before World War II broke out, he led one final expedition off the coast of Tunis to study the deep sea life of the Mediterranean.

While German soldiers occupied his country, Bruun made his plans for the day when peace would come. Working with the material from the *Dana* expedition, he became convinced that another such expedition was needed. In all its wanderings, the ship had never trawled in water deeper than 9,000 feet. There are, however, many places in the seas deeper than this. Such deeps equal in area more than half that of Europe. Could anything exist down there in such terrible pressures (equal to that of 1,000 atmospheres), such cold (only a few degrees above freezing), such darkness, and with such limited amounts of food available? Bruun thought there could but that they would be very strange creatures indeed.

As soon as the Germans moved out of Denmark, he began the complex job of organizing another round-the-world expedition. While gathering together the men, materials, and money needed, the impatient scientist, confined to his own country for five years, launched a comparatively short, modest trip.

PETER RASMUSSEN

*Anton Bruun, one of the great Danish ocean scientists*

A well-to-do Danish sculptor named Viggo Jarl offered to lend the Carlsberg Foundation his three-masted motor yacht for ten months and, in addition, to pay all expenses for a trip. In effect, this was an offer to Bruun, and he quickly accepted. He chose the west coast of Africa as his scene of operations partly because, at this time in 1945, the war in the Pacific was hardly over and conditions there unstable. The more important reason for the choice was that all the reports of great sea serpents over the years had come from this area, the southeast Atlantic. To find a great monster of the deep had become a consuming ambition of the Danish ichthyologist.

The yacht, the *Atlantide,* carried seventeen men. Anton Bruun, of course, was leader of the scientific team. To equip the vessel for research he borrowed an echo-sounding apparatus and an electric winch from the Danish Navy. For what was basically a fishing expedition, he also borrowed 12,000 scarce feet of wire from the British Navy and enlisted a colleague, F. C. Fraser, from the British Museum.

A number of things went wrong on this "little voyage" of 13,362 miles. The winch could only hold 5,600 feet of wire, and this determined the depth limit of their operations. There was only one cabin that could be converted into a lab, and it was just big enough to hold one man at a time. Most of the sorting and preserving of specimens had to be done out on deck in all kinds of weather. The engine could not operate at a speed of less than three knots, although trawling (deep-sea fishing) ideally should be done at speeds less than this. Off the Cape Verde Islands the ship's motorboat was lost, and for a time they had only one lifeboat, operated by sail. The *Atlantide* arrived off the Belgian Congo three weeks overdue because of numerous gales and a certain caution about operating in various mine fields left over from the recent war.

Nevertheless, the scientists harvested a considerable amount of data, the captain harpooned a dolphin, and Bruun was very delighted to be able to visit the beach at Dakar where he roamed, wearing his heavy leg brace, among the skeletons of 151 whales. In a very bizarre zoological event, a whole school of the huge animals had been stranded there three years before.

As almost always during expeditions, Bruun was a very happy man in spite of his disability. One day, as the *Atlantide* cruised along he was out on deck with the Englishman, Fraser, measuring and identifying fish. Turning to his friend, Bruun remarked, "Well, we may not be great zoologists but no one can deny that we are enjoying ourselves."

On returning to Denmark, he worked very diligently to put his round-the-world expedition together. This involved not only securing the necessary funds but also selecting the scientific subjects to be investigated, deciding which route the ship would take, and finding the proper ship. This turned out to be a vessel which the British Navy wanted to sell, as they were cutting back their fleet after the war. Built as a yacht for the Governor of Fiji, it saw, instead, service as a mine sweeper. Bruun, in the sentimental way of oceanographers, named the vessel he purchased *Galathea,* after a corvette in which a Danish explorer, Steen Bille, had sailed around the world from 1845 to 1847.

So that he might concentrate on his own special research interests, Bruun looked around for someone else to be the working leader of the scientific party. Such an idea was unthinkable, however. In a tribute to him, his student, Torben Wolff, said, "The Galathea Expedition stands as a milestone in the history of oceanography. This fact is in the first place due to its leader, Dr. Anton Bruun.

"Firstly, a profound knowledge, experience and common sense are invaluable resources for any leader of a scientific enterprise. I can hardly imagine anyone more qualified in the field of biological oceanography than Bruun.

"Secondly, he became from the first day the natural center on board. His well-balanced and quiet temper, his unending pleasure and ability to discuss all matters under the sun, his sense of justice as well as his sense of humor, and his faculty to say the right thing in the right place, made him equally loved and respected among the crew and the scientific staff. He held an essential part in the excellent spirit which prevailed on board the *Galathea.*

"Thirdly, he was the best imaginable representative and ambassador for the expedition and for marine sciences. His personal, legendary charm and his knowledge of languages, his interest in people of all kinds and in foreign cultures were among the most important ingredients in the establishment of the splendor which surrounded the vessel in one port after the other, throughout those two years."

A man who inspires this kind of admiration realizes, however modestly, that people are going to give him very little time to himself. Nevertheless, he did not abandon his personal goal. "The Great Sea Serpent is yet to be caught," he said, and ordered that the ship sail on October 15, 1950.

The *Galathea* sailed down the west coast of Africa, around the Cape of Good Hope, and into the Indian Ocean, pausing frequently to trawl where the water was particularly deep, and occasionally stopping at interesting ports to pay scientific social calls.

When trawling, the men on the *Galathea* used very modern equipment that the crew on the old *Dana* most certainly would have envied. As important as any thing was the echo-sounding gear, by means of which they located the narrow trenches, miles deep, which exist alongside certain arcs of islands. These long slashes on the ocean bottom had been inaccessible to the comparatively primitive tools Bruun had had on his first two-year trip around the world. Now, with the new depth-finding devices, once a trench had been found, a continuous echo could be obtained to make certain that the trawl was working as far down as possible. The trawl itself was an enormous net, sixty feet long with a huge mouth twenty feet wide, capable of capturing as much as four tons of sea life in a single haul. The net was lowered and raised by a powerful winch that carried a line of steel wire seven and one half miles in length.

To illustrate what the depths were like, Bruun pointed out that if Mount Everest were dropped into the Philippine Trench, its peak would be a mile and a half below sea level. "To drop a thin line into this fantastically deep chasm, to grope blindly for organisms is comparable to flying an airplane six miles up and trying to snag animals on the ground with a grappling hook from a mile-long drag line. . . . An airplane explorer who pulled up a mouse on his line might conclude that the earth's surface was largely inhabited by tiny rodents."

Bruun reasoned, however, that what the *Galathea* came up with was a reasonable sample of the types of beasts that lived in trenches and that life at such depths was probably quite similar in all parts of the world.

While sailing in the Indian Ocean, the *Galathea* stopped at Madagascar, Mombasa, and the Seychelles Islands. One of the world's most curious chains of islands, the Seychelles are built out of granite, a base which everywhere else on earth is the exclusive property of continents. The islands are also the only place in the world where *cocos-de-mer,* or sea coconuts, grow. These fifty-pound seeds, which sometimes wash ashore at far-away places, were once thought to have properties so magical that rich men in Southeast Asia would offer whole bags of pearls or an entire shipload of silk for a single specimen. Today they are just curiosities, but, none the less, the biggest nuts that grow anywhere.

When the *Galathea* arrived at Ceylon, Dr. Grace Pickford of Yale University stepped aboard. Scientists of many nations sailed with the ship during its cruise around the world, but Miss Pickford was the only woman. The well-fitted ship had proper accommodations for her. She had been invited because of her work as a biologist with deep-sea cephalopods, a class of marine life which includes squid, octopus, cuttlefish, and nautilus. Much of her study had been done in laboratories, and she never had the opportunity, for instance, to see a 'Vampyrotenthis' (a form of marine life that looks like a vampire bat) while still fresh from the deeps. When the ship stopped at Singapore for a few days, Miss Pickford impressed her male colleagues with her cool bravery—by putting her hands in all sorts of dark holes and crevices to find octopuses. Sometimes she needed assistance in removing their grasping tentacles from her arms.

Before being driven away by typhoons, the *Galathea* made its deepest trawls in the waters of the Philippine Trench. There, with a bottom sounding of 30,000 feet, 36,000 feet of line was released. The ship moved along the trench at a speed of two knots. When the net at last was winched to the surface, the scientists found clay on the trawl's framework, proof that it had touched bottom. Then they noticed a sea anemone attached to a stone in the net, evidence that the animal, too, had come from the bottom of the trench. At first it seemed typical of its species, but on close inspection it was found to be so different that it really belonged to an entirely new family. All in all, eight different species of animals were found in the Philippine Trench, showing that life in some form or another can accommodate itself to the most impossible conditions on earth.

At the middle levels of the sea, with a depth of around 6,000 feet, most of the animals are of the small, burrowing kind. They resemble relatives that live near the surface—except that, living in complete darkness, they are blind and usually colorless. Below a depth of about 18,000 feet, animals that have backbones almost disappear. The *Galathea,* with all its specimens, found only six, down that far, with vertebrae.

Animals that live in the very deep sea are safe from the usual predators. None of the carnivorous types can get to them because of the very special conditions down there. At the same time, once they have adapted to the tremendous pressures of the deep water in the trenches, the inhabitants cannot move out again. As on remote islands, they evolve into special forms, recognizably related to animals living at shallower levels, but at the same time considerably altered.

Many of these animals are simultaneously male and female. Called mouthbreeders, these animals hold their eggs orally until the young hatch and can care for themselves. Too little food exists three or four miles down to support great numbers of newborn, and no species would survive if fish wasted their larva in the spectacular way of surface dwellers who may lay a million eggs at a time.

The *Galathea* Expedition found life in the abysses to be much more abundant than scientists had previously believed. The dominant species turned out to be the holothurian, a sea cucumber having a soft body. On the most productive haul of the voyage the net contained 2,100 specimens, representing 132 various species. Anton Bruun's ability to recall instantly any type he had seen before became uniquely important when the specimens needed instant classification before they faded and decomposed on deck.

Scientifically, the expedition's most interesting find looks on one side little different from an ordinary clam. What makes it so odd is that when it is turned over, the other half-shell one expects on a clam is not there at all—the living muscle tissue of this creature has no protection such as modern clams and oysters have evolved. Until this expedition no one had ever seen such an animal alive. It was generally believed that an ancient mollusk like this, named Pilina, had become extinct some 350 million years ago. The men on the *Galathea* named their living fossil "Neopilina Galatheae."

Dr. Bruun wrote about another first-of-a-kind animal that they discovered. "We would occasionally feel exhilarated by the great catches and the advancement of our knowledge which we thought they gave; but then suddenly a creature would turn up to show that our achievement was but an inch on the way.

"Take 'Galatheahauma,' a creature with an appearance as formidable as the spelling of its name. It came up as late as May 6, 1952, in one of the concluding trawls of the expedition, off tropical West America [Central America]. It was from a depth of 3,590 meters (about 12,000 feet), 'only' 3,590 meters, as with our experience of five trenches exceeding 7,000 meters, we so grandly put it. Yet here was the most singular fish of the whole expedition—a black, broad-headed angler fish; and to my colleagues and my own delight I had to admit I had seen nothing like it on the *Dana*.

"The fact is that the *Dana* material includes the world's largest collection of deep-sea angler fishes. But our 'Galatheahauma' is not in it. We had caught a number of angler fishes with a light organ on a

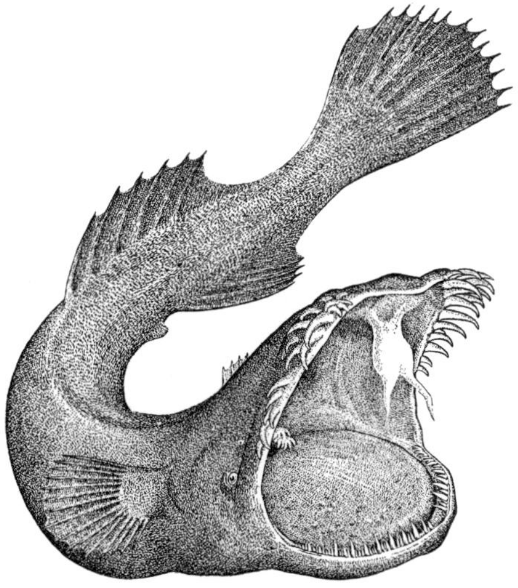

Galatheahauma, *one of the world's strangest underwater creatures*

stalk on the nose; we had also caught a number of these old acquaintances from the *Dana,* this time in the Panama region.

"Here we had an altogether different type. The large light organ is now *inside* the mouth, suspended from the roof behind pointed, curved teeth which fringe the upper jaw like a comb. We found later that a small specimen, eight centimeters long (about two-and-a-half inches) had been taken by an American expedition in 1908, in Indonesia in 1,385 meters of water and in a trawl which, like ours, had fished on the bottom. Compared with this and most other deep-sea angler fishes, our 'Galatheahauma' was a giant of 47 meters (almost 15 inches) and so different from any other that we were justified in giving it the new name.

"It seems to me probable that we have here a deep-sea angler fish which lives close to or on the bottom. There it need only lie with its jaws open, leaving the large light organ with its two fine extensions to lure the fish or prawn to it; then as soon as the prey is within reach of the long teeth the jaws will shut, the corners of the upper jaw falling down and trapping the prey against the lower jaw.

"This 'living mousetrap' with bait is unquestionably the strangest catch of the Galathea Expedition, and altogether one of the oddest creatures in the teeming variety of the fish world. But we caught only one, right at the end of the cruise, and still no one has caught the Great Sea Serpent."

Like many travelers, the *Galathea* had unforeseen expenses, mostly due to the high cost of everything in the Pacific, a situation brought on by the Korean War. Funds began to run low. The ship sailed through the Panama Canal three months early and returned to Copenhagen and a joyous reception on July 17, 1952. The scientists had nevertheless been gone for 19 months, made 782 stations (pauses to collect data), and traveled more than 63,000 miles. No one felt disappointed. No ship had ever done so much research in biology at such colossal depths, nor produced more results.

During the years that followed his great expedition, Anton Bruun gathered many honors and world-wide fame. The voyage had fulfilled his ambition, but the burdens of being a leader, particularly difficult for a man so handicapped, had cost him a good deal in strength and health.

As a result of the Galathea Expedition, Bruun became convinced that no one nation could learn all the secrets of the sea by itself. A little nation such as his Denmark could only afford, at best, one *Galathea* at a time. He realized that the sea was too vast, too complex, too unknown; that to make it as well understood as the land, required the combined efforts of all interested nations.

To help bring this about he became one of the original members of the International Oceanographic Commission of the United Nations. To promote scientific cooperation between countries he visited scientific meetings in places as separate as Peru, the Philippines, and the Soviet Union. He was one of the early supporters of the International Indian Ocean Expedition, which in 1963 and 1964 involved twenty-three nations and more than forty ships (as well as a number of airplanes) in exploration of the least known of the world's great bodies of water.

His interest in the food problems of the less developed countries led him not only to promoting a program for the Indian Ocean but also to accepting a visiting professorship from the Scripps Institution in California. Scripps had also invited him to lead an expedition across the Pacific with the purpose of promoting the fisheries of Southeast Asia and to train marine biologists and oceanographers in that part of the world.

None of this caused him to forget his own favorite topic. While planning to sail for Asia, he wrote, "I have never seen a Giant Sea Serpent but I have a very good reason to believe that such a creature exists. I saw an eel-like fish in tadpole stage six feet long, while a young zoologist aboard a Danish trawler off West Africa. There was a larval form of the eel in our nets. It was typical in all respects except size. It could be expected to grow to 72 feet. We weren't lucky enough to find his parents. I have searched for them for twenty years."

Bruun added that every report of giant creatures had been from the South Atlantic. In 1848 the entire crew of the frigate *Deadelus* saw one 60 feet long. The U.S.S. *Daphne* a few years later fired at one 100 feet in size in about the same area. Two zoologists in a boat off Brazil reported a serpent in 1905. A ship of the Grace Line ran over one in 1947, cutting it in two, again near Brazil. One side was 35 feet long.

If sea serpents do exist, the west coast of Africa, one of the world's richest feeding grounds, would be a natural home. Bruun wrote that he wanted to go back there some day and drop deep lines, with frozen herring or squid as bait.

He never did return, however. Ill health even prevented him from taking a very active part in the Scripps expedition to the South Pacific. He died in 1961 while in Copenhagen, "and this happened as he would have wished," his English friend, Fraser, wrote, "while discussing a thesis with his colleagues from the *Galathea,* and other marine zoologists, on the distribution and origin of abyssal fauna."

Among the many tributes to Bruun to reach his colleagues in Denmark was one from the scientists of the Soviet research ship *Vityaz:* "His role was especially great in the complex task of uniting the efforts of oceanographers of various countries and promoting the global study of the World Ocean."

The bronze plaque fixed to the wheelhouse of the U.S. research vessel which bore his name read, "To the memory of Anton Bruun, 1901-1961. Explorer of the sea, pioneer of the deeps, organizer and leader of expeditions and global programs, scientific statesman, citizen of the world and friend and inspiration of a generation of oceanographers—this ship is dedicated. May she bear his name proudly and contribute in full measure to the advancement of man's knowledge of the wide seas and their inhabitants."

For four years the *Anton Bruun* sailed the Indian Ocean and the South Pacific, primarily conducting biological research. Under the auspices of the National Science Foundation, the ship spent more time in the Indian Ocean than any other ship from the Americas or Europe, conducting the kind of biological research that Bruun did so much to promote. Finally, in 1967, it was retired from active duty. Newer, better equipped vessels had made the handsome vessel obsolete.

Trawling of the great deeps of the ocean will, nevertheless, continue. It may be hoped that when the Great Sea Serpent is finally found, it will be named in some Latin version of the name Anton Bruun.

# 5 Maurice Ewing

## *The Man Who Makes Earthquakes*

*In the pursuit of profits today, business often enlists the aid of science. Sometimes it works the other way, and a technique that is worked out to produce wealth may also prove to be very useful in the pursuit of basic knowledge. One man who achieved international fame as a scientist and leader of a great oceanographic institution learned his first lessons about the earth's crust while working for an oil company.*

THE MOST visible symbol of the many, many achievements of Dr. Maurice Ewing is a geological observatory hidden off to one side of busy Highway 9W at the state line between New Jersey and Palisades, New York. A division of Columbia University, Dr. Ewing's center for research into the history of the earth is named after Thomas Lamont, the partner of J. P. Morgan, whose widow gave her large weekend estate to the school. High above the Hudson River, the Lamont Geological Observatory rests on volcanic diabase rock that dates from the Triassic era about 170 million years ago—the same gray-red rock of the Jersey cliffs visible to motorists who cross the George Washington Bridge about fifteen miles to the south. Ordinarily one might not think about the ancient past of a site on which a campus of higher learning has been built, but the thought comes naturally at Lamont because the crust of our planet is the major concern of Dr. Ewing's institution. Since its establishment in 1949, the Lamont Observatory has added considerably to our knowledge of the planet and, because of necessity, to the sciences of oceanography.

Oceanography has become a necessary study there because about three-fourths of the world is covered by water. This fact may irritate men who would find it easier to study the earth's surface and what is underneath, with pick and shovel and their own eyesight. Since it is not yet possible to do much looking around under two or three miles of water, Dr. Ewing in his first major scientific accomplishment devised a new way of using instruments as substitutes for direct observation.

Maurice Ewing, who today stands six feet two and whose head is covered with unruly white hair, was born in a little Texas town far from the sea. His father was a farmer and a merchant. Ewing, who became an oceanographer because the sea was in his way, is one of a family of seven children. He earned his bachelor's, master's and doctor's degrees at Rice Institute in Houston, where he specialized in mathematics and physics. While a student, he spent summer vacations working with oil company teams prospecting off the shores of Texas and Louisiana. The technique the teams used to find salt domes, curvatures in the earth which are the likely locations of underwater oil wells, was to drop explosive charges over the side of a ship and from the recorded echoes get a picture of the shape of the ocean bottom. "If you want to sum up what I have done," Dr. Ewing says now, "you can say I have adapted the tools and tricks of the oil companies to exploration of the deep sea."

This did not happen immediately, of course. First he went to the University of Pittsburgh as an instructor in physics and then in 1930 to Lehigh University. "I was deeply interested in geophysical prospecting, a holdover from my student days. At Lehigh we made some seismological observations of quarry blasting and published some papers on the studies."

Two distinguished scientists, one at Princeton and the other at the U.S. Coast and Geodetic Survey, read these papers and asked Ewing, who was then in his twenties, if he would like to try this same kind of surveying out in the Atlantic on the Continental Shelf. "Little was known of the geological structure of this area of transition from continent to ocean. If I could get seismic readings of it comparable to those I had been getting on land, the information might provide an approach to the fundamental question, 'Why are some parts of the earth continents and other parts ocean?' I was so desperate for a chance to do research that if they had said the moon, rather than the ocean, I would have tackled it."

So, with two assistants, Dr. Ewing carried a portable seismograph, the instrument that was designed to detect earthquakes, across the coastal plain of Virginia from Petersburg to Cape Henry and then aboard the Coast and Geodetic Survey vessel *Oceanographer*. Every

LAMONT GEOLOGICAL OBSERVATORY, COLUMBIA UNIVERSITY

*Maurice Ewing, founder of the Lamont Geological Observatory and leader of many deep-sea expeditions*

few miles as the ship proceeded eastward it was stopped and the seismograph lowered to the sea floor. Then the three young men would set off in a whaleboat, sink blasting charges to the ocean bottom, and then explode them. As the charges went off the seismograph picked up the waves of energy that were reflected and refracted from the earth under the sea bottom. The difference in the amount of time the waves took to return from the surface bottom (the top of the sediment) and the true ocean floor beneath the sediment (the mantle), told the thickness of the sediment layer and gave an idea of its density. A little later, Dr. Ewing was invited to repeat the same kind of experiments while aboard the *Atlantis,* the Woods Hole Institution's research sailing ship. Some scientists thought that the sediment might measure as much as 12 miles in depth, while others thought it might be only 500 feet thick. In the area he studied, Maurice Ewing discovered that the refracted waves gave a thickness that varied from 1,500 to 3,000 feet.

Since those early days in 1936, seismic shooting has been much refined. It is no longer necessary to lay an instrument on the ocean floor nor explode the charges of TNT on the bottom. The energy waves from the explosions are now picked up by hydrophones, usually on a companion ship that may be 50 or 60 miles away. The waves are generated by depth charges set to explode at planned depths or, more recently, by electric sparkers which are a good deal safer to use than explosives.

Seismic shooting has become a major activity in oceanography. These tiny, man-made earthquakes occur frequently now in all the oceans, including the Arctic and Antarctic. The *Vema,* one of the two ships that Lamont now operates, has made more than 30,000 miles of seismic profiles of the sea bottom.

It has been discovered that oceans and continents differ in more than just their elevation. They are quite different kinds of structures. Now it is known that the rocks under the ocean are made of basalt and not of granite, as they are under the continents. Seismic shooting has also shown, contrary to former beliefs, that the Pacific is not the oldest ocean and that its crust and that of the Atlantic are very much the same. Another finding is that the crust under both oceans has a thickness which averages 3 miles down to the mantle, compared to the 15 to 24 mile-deep crust found under the continents.

Although Dr. Ewing had become an officer in the U.S. Naval Reserve, when World War II seemed imminent he was assigned to the Woods Hole Oceanographic Institution in Massachusetts, but as a civilian. The Navy had become actively interested in the oddities of sound propagation in the sea, and the scientists at Woods Hole knew more about these phenomena than anyone else. Dr. Ewing was to help the continuing research on the subject.

As a hobby he had been developing a camera for deep-sea work that would picture what the bottom looked like and show much more accurately than would lowering blind objects what life there was in the sea. He told the Navy about his camera and was encouraged to perfect it for use in photographing the sunken wrecks of ships.

Then in 1944, the newspapers announced that Dr. Maurice Ewing had been appointed Associate Professor of Geophysics at Columbia University, a post to be taken up at the war's end. At the same time it was said that "Dr. Ewing's studies of the ocean floor had proved of value to the war effort." Just what these studies were, however, was kept secret at the time.

Dr. Ewing's contribution was, in fact, a method adapted from his seismic shooting by which men from ships and planes lost at sea could be rescued. What the survivor was supposed to do was to drop a depth charge set to explode at certain depths between 3,000 and 4,000 feet. This would create underwater sound waves that could be detected by hydrophones at shore stations. Survivors could be located within a square mile from a station 2,000 miles away. Asked how long it would take for a survivor to be rescued, Dr. Ewing replied that it would depend on his position. "In the Pacific I would estimate roughly the maximum time required for rescue as around eight hours. Compare that with the twenty-two days it took to find Eddie Rickenbacker." (This famous Air Force General had been shot down in the Pacific during the war by the Japanese.)

This rescue system sounds ideal but in practice contains some difficulties. What it reveals without quite meaning to is the militarily astonishing fact that Ewing had shown the Navy how to pin-point a source of sound energy thousands of miles away. For undersea warfare this is a very useful thing to know how to do. In a citation the Navy called his work "the most specular discovery made during the war in the field of sound physics."

When Dr. Ewing went to Columbia University he was given two small rooms in the basement of a building on Morningside Heights in New York City as his office and laboratory. What with the surface traffic above and the rumble of the Broadway subway below, sensitive seismographic work was an impossibility. His physical research work was confined to the cruises he took in summer aboard the Woods Hole's *Atlantis.* Then Mrs. Lamont offered her large house and her one hundred acre estate to the University, and Dr. Ewing's problem was solved. In 1949 he was named director of the newly-formed observatory and appointed Higgins Professor of Geology, although he had never taken a course in geology in his life. He says, "It took guts to do that."

Today, Lamont is one of the United States' greatest centers for oceanographic research. With substantial grants from the U.S. Government, it now has a staff of four hundred people, seventy graduate students, fourteen buildings, and two ships. One, the *Conrad,* is operated for the U.S. Navy. The other, the *Vema,* a 202-foot, 500-ton schooner, was bought for Lamont in 1953. Every year since then, Dr. Ewing has been aboard it as scientific director of its oceanographic cruises. The major research effort has always been into the nature of the earth's crust. But other interests have included the study of earthquakes, measuring the age of sea water from the radioactive isotopes it contains, determining the world's past weather history (paleoclimatology), investigating the planet's past through samples of the ocean sediments (micropaleontology), and the study of the sea's abyssal plains, turbidity currents (ocean landslides), and the Mid-Ocean Ridge, which circles the earth two and one-half times.

The little *Vema* has cruised in every ocean and logged many hundreds of thousands of miles but, naturally, has spent more time in the nearby North Atlantic. There, in January of the year after Lamont acquired the ship, the Observatory almost lost its director. While the ship was sailing 200 miles north of Bermuda, a sudden storm came up and four men from the crew of thirteen were swept overboard. Among those cast into the frigid rough sea were Maurice Ewing and his brother, John. The sailing ship maneuvered as best it could to rescue them, but vessels like this are very difficult to handle in such weather. As he swam, Dr. Ewing heard the mate, Charles Wilkie, cry out, and then Ewing, helplessly watched him sink under the waves. Then the fourth man, Mike Brown, shouted. He was hanging on to an oil drum that had also been swept overboard. Ewing managed to swim over and join him.

As they hung on, they watched the *Vema* pick up John Ewing and then come to their aid. When their turn came and they were finally aboard, Dr. Ewing collapsed. As he said in a letter to his children, "At forty-seven I am too old for such heavy swimming. But love for you all gave me the strength to keep afloat long after I was exhausted."

The accident, naturally, did not interfere with Maurice Ewing's further investigations at sea, nor on land. As the Observatory's annual bulletin once said, "Almost all the detailed information on the structure of the earth's deep interior has been and probably will continue to be derived from the study of propagation of seismic waves." Thus the work in man-made and natural earthquakes continued. Elaborate seismic listening stations, without parallel elsewhere, are operated there at Palisades and in a deep mine near by in Ogdensburg, New Jersey. Other special instruments are operated in cooperation with local institutions at fifteen other sites around the world. Each seismograph is tuned to a different wave length in the earthquake spectrum, and among them, everything from a tiny micro-seism started by a storm at sea to the great quake that occurred in Alaska in 1964 can be detected.

One complete seismic observatory from Lamont was placed on the bottom of the Pacific, 100 miles seaward from Northern California. A similar instrument has been put down in 15,000 feet of water near Bermuda. By remote control the Pacific instrument can carry out thirty-five different commands. It is one hundred times more sensitive to disturbances of the earth's crust than those which are used on land.

These two special seismographs are sponsored by the U.S. Navy because they can be used to detect nuclear explosions and because they can spot disturbances of the earth's crust caused by waves, and thus be used as storm warning stations.

These two seismographs are also being used to investigate a still unexplained phenomenon. Some types of earthquake waves which originate in the ocean cannot be detected on shore-based instruments. Other earthquakes, happening on land, cannot be spotted on ocean-based instruments. If there proves to be a real barrier to the passage of earthquake waves between the land and sea areas of the planet, it may turn out that not only are the continents and oceans distinctly divided but that the earth's mantle has distinct land and ocean divisions as well.

Another method of studying the ocean bottom used by Dr. Ewing, as by oceanographers everywhere, is to punch holes in it. This is called coring. What happens is that a steel tube called a corer, two and a half inches in diameter and as much as 70 feet long, is lowered to within 15 feet of the sedimentary bottom. At this point, a trigger trips a weight which drives the corer down into the ground.

Sediments which have been accumulating undisturbed perhaps for millions of years are brought up inside the corer to the surface where they can be analyzed. Microscopic bits of organic matter can be analyzed visually, chemically, and in terms of radioactivity to determine their age and to add to the knowledge of climatic changes in the earth's past.

One curious fact discovered about the sediments is that so far none more than 100 million years old have been found. This is odd because many sedimentary rocks on the continents are quite a few times more than that age. It seems to indicate that the ocean floor went through some tremendous upheaval around 100 million years ago and eliminated all traces of previous history.

One problem with sediments is that they vary considerably in thickness and, especially in parts of the Pacific, are far shallower than scientists would reasonably expect.

Speaking of this kind of work, Dr. Ewing has said, "We are steadily increasing the length of our core samples. The seismological data indicate that the thickness of the sedimentary layer on most of the ocean floor is only about 2,000 feet. We have every reason to believe that in that 2,000 feet of sediment the whole history of the earth is better preserved than it is in the continental rocks . . . the dream of my life is to punch that hole 2,000 feet deep and bring the contents back to the lab and study them."

To get samples of this length, scientists are turning to drilling. Working with several other institutions of oceanography and with a commercial drilling ship, men from Lamont have drilled more than 1,000 feet into the sediment in deep water off the coast of Florida. This is the farthest anyone has ever drilled into the ocean for scientific reasons. The samples, which take years to study, are still being analyzed. (Parenthetically, this drilling produced a fountain of fresh water, far out to sea, that spurted 30 feet into the air. The oceanographers had found a new source of water for Jacksonville.)

The ultimate in drilling was the Mohole project, which Dr. Ewing originally approved of as a member of the first advisory board. The idea was to bore a hole right through the crust of the earth where it was thinnest at the bottom of the sea, about three miles down to the mysterious mantle of the earth. The discontinuity between the crust and the mantle, named after the Yugoslav seismologist, Andrija Mohorovicic, who first discovered it while studying waves from earthquakes, is so far known only theoretically. If the Mohole project, which Congress discontinued in 1966 because of the expense, should be revived, and be successful, it would add a tremendous amount to the world's sum of knowledge. For the present, Dr. Ewing has withdrawn his support of the Mohole, but not because of politics. He feels that the risk of failure is too great, and that science needs to know a great deal more about the earth before attempting it.

Maurice Ewing's curiosity has led him not only to study what is under the sea but the shape of the sea bottom itself. While still sailing in the *Atlantis* in 1947 and 1948, he made two epic voyages to the Mid-Atlantic Ridge, the great range of mountains that runs down to the center of the ocean between Europe and North America. During these voyages Bermuda was found to be the top of a huge mountain. A seamount the size of the Matterhorn was also discovered. A trawl sent two miles deep in the water found 200 different species of sea life. A core taken on a seamount midway across the Atlantic produced sediment that was deposited 60 million years ago. This was at a depth of a mile.

Dr. Ewing believes that finding such an old sediment on a seamount seems to disprove the theory of Continental Drift—the idea that the Americas, Europe, and Africa were once joined together but somehow drifted apart—because the seamount could not exist if the space for it to grow in did not exist. On the other hand, one might say that this sediment only proves that the drift, it it happened, occurred more than 60 million years ago.

On this same voyage, Ewing and his colleagues found that the Hudson Canyon, off the coast of the United States near New York, is 100 miles longer than had been supposed. If the theory is correct that this gash in the sea bottom was cut by the Hudson River, as the Grand Canyon has been formed by the Colorado, then either the sea level at some time was two miles lower than it is today or the U.S. coast has risen by the same amount. Dr. Ewing accounts for the existence of the Hudson Canyon with the explanation that it was cut by turbidity currents. These are giant landslides of sediment that sometimes tumble off the Continental Shelf and pour down the steep cliffs into the abyss below.

As Dr. Ewing and his associates began to analyze the soundings of these expeditions and reports about the same area from other research groups, the pattern of the Atlantic bottom began to take on shape. Part of the story at Lamont concerns Marie Tharp, a geophysicist who was making a diagram of what was then known about the floor of the Atlantic Ocean. At that time Miss Tharp did not have a tremendous number of soundings but still enough to make her wonder if she did not see something no one had observed before. In the already known Mid-Atlantic Ridge that runs between Iceland and the Azores, the depth figures available seemed to show a valley right down the central spine. She brought the idea to the attention of Bruce Heezen, one of Dr. Ewing's associates, who agreed that it was possible but that there was hardly enough information to get excited about.

Nothing more was said about the valley until a telephone company came to Bruce Heezen for help in understanding an unusual number of breaks in their trans-Atlantic cable. The cause turned out to be earthquakes. Miss Tharp, Dr. Heezen, and Dr. Ewing studied the matter for five years before writing a joint paper on the subject. In it they said that under the sea there exists a mid-ocean mountain range, 40,000 miles long, that circles the earth two-and-a-half times, of which the Mid-Atlantic Ridge is a part. More startling, their paper reported a rift valley which appears everywhere along the center of the range. This announcement surprised the scientific community at the time, but today the idea seems beyond dispute.

Dr. Heezen feels that this long rift valley is caused by the earth's attempt to expand. Matter deep inside the earth, heated by radioactivity, bubbles away like thick soup, expands, and pushes against the earth's surface, causing a rift. This theory would explain the numerous earthquakes in the rift and the number of volcanoes which occur along it. Later, Dr. Roger Revelle and his associates at Scripps conducted experiments in the Pacific which led them to also propose that as the earth expands it is splitting apart.

For some years Dr. Ewing has been extremely interested in the earth's several Ice Ages. What caused them and when did they happen? From his study of sediments he knew that a very abrupt change of climate occurred 11,000 years ago. That was the end of the last Ice Age.

A theory which he and William Donn have revised slightly over the years is that the Ice Ages began when the planet's axis shifted. Less sunlight in the middle and higher latitudes caused a drop in temperature. This created the zones of climate, from the heat of the Equator to the cold at the poles. In the northern hemisphere, the continental ice sheet began to grow in polar latitudes with snow from the ocean itself, which became frozen. The ice sheet was also fed by snow from the south.

When the glacier reached its height, sea level had fallen so low that there was little evaporation, thus little precipitation. For lack of moisture, the ice sheets began to wither away. The next Ice Age, Ewing and Donn believe, will start when the north polar ice pack disappears. Evidence shows, in fact, that it has been warming up since 1900. Like many another great problem in science, Ewing's and Donn's explanation of the Ice Ages satisfies some people but by no means all of them.

Maurice Ewing is a theorist, a practicing geophysicist, a sailor, and administrator. He developed the first camera able to photograph at great depths and has made improved echo sounders, magnetometers, bathythermographs, core and sediment temperature probes. He is also a prolific author and has written three books and hundreds of papers. Not designed as light reading, they cover such subjects as earthquake seismology, submarine geology, gravity surveys, submarine acoustics, seismic refraction, climatology, and geothermal measuring. Numerous medals and honors have been awarded to him for his work.

At the Lamont Geological Observatory, of which he has been in charge since its first day, research and development continue to go on in many directions. A seismograph, for instance, is being created for emplacement one day upon the surface of the moon. Current excitement at the institution centers on magnetic studies of the ocean floor. Many cores of sediment show that the north and south magnetic poles have reversed their fields a number of times. This may explain the complete destruction of various species of life, such as the dinosaurs, that have occurred from time to time. The earth's magnetism provides a shield which cosmic rays cannot penetrate. But without this magnetism during a period of reversal the rays could easily kill whole races of animals and plants and create new ones. The suggestion has been made at Lamont that a reversal of magnetism may occur when a large object enters the earth's atmosphere and explodes. This reversal seems to have occurred at least once, about 700,000 years ago, when something about a mile in diameter exploded near the earth and sprayed Japan, Australia, and Africa with glassy fragments known as tecktites. These are found on land, in "strewn fields," but occur frequently in ocean sediments.

As time goes by, other great theories will have their day to be finally proved or perhaps to be found wrong in some way. In any case, the search for truth will continue at the dynamic institution which Dr. Ewing founded.

He himself will continue to work on such questions as why there are continents and how their granite got there, as well as "the grand problem—which is the structure, composition, limiting conditions, and history of the planet that is our home in the universe."

# 6 Edwin A. Link

## *The Limits of Human Experience*

*Some men are not satisfied with success in one field. An innovator who has seen one brain child grow to maturity may restlessly move on to another, not seeking more fame but the solutions to new problems. He may also be of a cast of mind which sees that research and development, and adventure, may just as well be good business at the same time.*

FOR THREE HUNDRED years nothing but fish had lived in the town of Port Royal, Jamaica, until Edwin A. Link entered it from his yacht *Sea Diver*. During the time of England's William and Mary, Jamaica had been part of the Empire, but this colony's principal city was one of the least respectable gems in the royal crown. Port Royal served primarily as a hideout and trading port for pirates.

Jamaica, like all the islands in the West Indies, keeps at best only an uneasy truce with nature. Unlike other islands in the Caribbean arc, it does not have active volcanoes but hurricanes must figure in every Jamaican's reckoning. Earthquakes, however, are very uncommon. Nevertheless, just before noon on June 7, 1692, the bustling, lawless town of Port Royal was rudely dumped into the sea. A part of the earth's crust had slipped.

The community vanished under the waves with a terrible loss of life. Over the centuries Jamaican natives told stories of hearing submerged church bells ringing. But though the location of the disaster was well known, no one had visited it until Link, in his capacity of underwater archeologist, arrived there in 1959.

With his wife, Marian, and members of his well-trained crew, the man who had gained his fame as the inventor of an aircraft trainer went down in diving gear to inspect the ruins. The water was murky, the rubble from the buildings overgrown with sea life and strewn about in confusion, but the party had studied an old plan of the town and knew where the most interesting ruins might be found.

The divers came upon warehouses, homes, and the remains of what must have been the kitchen of a restaurant. They found ancient bottles, pots and pans, and evidence that the noon meal was just being prepared when the earthquake struck without warning. Among the many items the divers brought to the surface was a brass case. Opening it, they found a pocket watch. The name "Paul Blondel" could just be deciphered on the face. The hands of the watch told the time when it stopped running: seventeen minutes before twelve, the exact moment of the earthquake. Link's later research in Europe showed that there had been a watchmaker of that name in the Netherlands before 1686. Today the watch rests in a museum at Kingston, Jamaica.

Such undersea activities came fairly late in the life of Mr. Link although as a boy growing up in Binghampton, New York, he had learned a considerable amount about boating on the nearby Finger Lakes. His ambition was to be an inventor. With this objective, his early studies were in engineering. Aviation was still very new when he was young, and fascinated with the idea of flying, he earned his pilot's license shortly after he was old enough to vote. And it was through his work in the field of aviation that he gained the wealth which made it possible for him to take a yacht to Jamaica.

Edwin Link has nineteen patents to his name, but the one for which he is most generally known is the Link Flight Trainer and Simulator. This ingenious device can put a novice pilot through almost all the stages of flying while he is still safely on the ground. Starting before World War II, thousands and thousands of flyers have learned their skills aboard the celebrated Link Trainer.

Link has published a number of highly technical papers on the subject of navigation, but his major achievement has been to make it easier for less abstract minds to find their way about in the sky. Also concerned with history, Link studied the career of Christopher Columbus and hailed him as the "man who simplified celestial navigation." Link's first intellectual adventure with the sea, a journey which his wife Marian shared, was to determine where Columbus had actually sailed and to find out at which island the discoverer of the Americas made his first landfall.

This concern with Columbus became the turning point in a new life for the inventor. With his Trainer firmly established and his business prospering, Link decided to retire from full-time occupation with it and to pit his active mind against the problems of the sea.

OCEAN SYSTEMS, INC.

Deep Diver, *the underwater craft developed by Edwin Link to put men in the sea, being lowered into the water*

He began with underwater archeology. In Florida he had the 91-foot *Sea Diver* specially built for research work, and it went into commission in 1959. Exploring Port Royal was his first project. The following year he took the ship to Palestine, where he and his men made dives down to the ancient sunken harbor of Caesarea, Herod's capital. They found some treasure there but even more when they took the diving gear overland and into the biblical Sea of Galilee. Discoveries included an anchor made of basalt, several pots, and a precious hoard of ancient glass.

In the summer of 1961 the *Sea Diver* was off the coast of Greece. There the Link-party explored an old Temple of Hera, now deep in the sea. While at this work, Link was suddenly arrested by officials of the Greek government. The explorer believed he had received all the necessary permissions, but the official charge was that he was making "unauthorized dives." The fine came to $740. Thoroughly irritated, Link announced to reporters that he was "leaving the country at once."

Governments were only one problem the Link team found down under the sea. A more permanent concern was that the divers had so little time in the water in which to do useful work.

A man who spends one hour on compressed air in 300 feet of water must spend seven and a half hours in a decompression chamber.

To Link, as to Captain George Bond and Jacques Yves Cousteau, this seemed terribly wasteful. Independently of the other two men but making use of Bond's research in exotic gases, Link designed his own deep-sea housing equipment with the idea of living for weeks or even months at the bottom of the ocean. In 1962, when he presented his project to a scientific committee who were considering sponsoring it, Link was received with considerable scepticism.

Nevertheless, the equipment was ready the summer of 1963 and he took it aboard the *Sea Diver* to Monaco. With him were his wife and his twenty-one year old son, Clayton, an experienced diver. The actual scene of operations was nearby Villefranche Bay, partially sheltered, and which had a number of different depths useful for the experiments. The bay was also headquarters for the U.S. Sixth Fleet. The Navy offered its help, if necessary.

Link had built a submersible pressure chamber from which a man could move out into the sea from depths as great as 400 feet. He would be able to return to the cylinder for rest and warmth whenever he needed to. Decompression would still be necessary, of course, but it would only need to be done once, no matter how long a man stayed down. The chamber had all the conveniences—heat, air conditioning, electric light, and a telephone. The breathing mixture was 97% helium and 3% oxygen.

Although he was fifty-eight years old, the short, sturdy inventor insisted that he himself must first test the safety of his own invention. At nine o'clock one morning late in August he went over the side, wearing a black rubber suit. He swam to the cylinder waiting for him in the water and dove down to the hatch that had been opened by an assistant. Entering, he closed it up and winched the cylinder down to 60 feet, along a chain that had been anchored to the ocean bottom. By means of controls aboard the *Sea Diver,* pressure within the cylinder was equalized with pressure in the water at that depth. Link emerged from the cylinder, breathing through a rubber hose which connected his face mask to a tank inside the housing.

After a time out in the water, he returned to his chamber to be met there by his son, Clayton, who had swum down using his aqualung, bringing the morning mail sealed in a jar. Link dictated some urgent replies over the telephone, a stunt made difficult for the secretary on the surface due to the helium, which makes the human voice sound like that of a duck.

About one o'clock in the afternoon, Clayton swam down again, carrying his father's lunch. Link spent most of the afternoon roaming about the sea. At 5:35 he had spent eight hours, in and out of the capsule, at a depth of 60 feet. This was the goal he had set himself. He closed his hatches and gradually winched the cylinder to the surface, not arriving until nine in the evening. Meanwhile, the pressure inside the cylinder was being slowly brought back to that of sea level. When the capsule surfaced, it was hoisted aboard the *Sea Diver,* but Link was not finally decompressed until 11:35 that night. During this time he had a good sleep. Finally, he emerged after fourteen hours in the chamber, laughing and in excellent spirits after the successful trial.

Link conceded that the next, more rigorous experiment, needed a younger man. He was Robert Stenuit, a diver from Belgium with a great deal of experience. As planned, Stenuit would live and work at a depth of 200 feet for at least a full day, possibly two or three. His trial began on the morning of September 6. He reached his depth at twelve noon. All went well for the first twenty hours—except that the people at the surface could not understand his speech, and the water outside was so cold that Stenuit could not stay out in it for more than a few minutes at a time, Stenuit slept during the night, though somewhat uncomfortably, in his cramped quarters. At seven in the morning he signaled "All's well" by means of Morse code. He added that he was planning to spend two more days down there.

All was not well, however. Watchers on the *Sea Diver* had just discovered that a great deal of helium had been wasted in the night because of leaking lines. There was no alarm yet. According to the decompression tables, a safe margin of gas still remained in the bottles.

Nevertheless, a launch was sent ashore to obtain reserve supplies. Then, as the boat sped back from Villefranche with fifteen bottles weighing more than a ton, a quick Mediterranean gale sprang up. The overloaded vessel began to take on water. All the bottles were in the stern, and the launch began to go down rapidly before the men could shift some of the bottles forward. The launch sank in a few minutes with all its valuable cargo. The men, of course, were saved.

Now there was a real emergency. The helium supply had dwindled. There seemed to be just enough to clear Stenuit's blood of nitrogen if he came to the surface immediately. Not knowing what had happened, he objected strenuously when told he was being brought up as quickly as possible. Once up, however, the weather was so rough that it seemed safer to leave him in the water than to try to swing the capsule aboard. Decompression had begun, however. After several hours, during a lull, the diver and his housing were at last recovered.

Stenuit had lived down in 200 feet of water for twenty-four hours and fifteen minutes. A depth record was set, and it could obviously have been exceeded but for the accident with the launch.

The following April Ed Link was once more in the Mediterranean with the *Sea Diver.* Returning from a cruise, he had just moored his ship in Monaco when he saw a high-ranking U.S. naval officer hurrying down the dock. Would Ed please phone the head of the U.S. Naval Oceanographic Office in Washington immediately?

While waiting for the call to go through, he made a guess as to what it was all about. Aboard the *Sea Diver* he had heard radio reports of the loss of the *Thresher,* a U.S. nuclear powered submarine. On the air every day there came new accounts of efforts to find the ship. When the voice finally came through from Washington, he discovered that the matter did indeed concern the *Thresher.* The Navy was shocked to discover that it had no means of helping the stricken sub nor even a way of finding the wreck. The Oceanographic Office had been given the job of correcting this situation. Would Link come to Washington and give them his advice on working in deep water?

Link had been planning to take the *Sea Diver* to Bermuda in a few days to continue deep-diving experiments in warmer water. It would take three weeks to get there, and Link had no one but himself to pilot the ship. Now he was being asked to drop all his plans and fly to Washington immediately. It is an honor to have the government ask you for advice—but it can also be a bit irritating.

Of course Link did agree to go, at least for a preliminary talk. Once in Washington, he became so fascinated with the possibilities of a major government-financed program for work in the depths that he shelved his own projects for many months. Although Link realized that this work initially would be for the benefit of the Navy, he foresaw how it could be applied to many civilian problems.

He became part of the Deep Submergence Systems Group, also known as the DSSRG. First, they considered the situation that was made obvious when the *Thresher* sank without a trace. Since the 1930's there had been no new developments in submarine salvage or rescue techniques. Nuclear submarines were operating far below the depths in which anyone could help them if there were trouble. Even if the *Thresher* had sunk in depths that would not have crushed her, there would still have been no way to rescue the crew. It had taken weeks even before the sunken vessel could be located. The DSSRG group recommended that in the future all submarines be equipped with a sonar that would respond automatically to a sonar on a search ship and also with a distress buoy that would rise quickly to the surface and radio an SOS.

Once the *Thresher* was located, the problem of rescuing remained. At the time of the disaster, the best tool the Navy had was an eight-man diving bell, lowered from a surface vessel, that often had to be sealed to the distressed sub's escape hatch by divers. Since divers could only go down to 500 feet (the *Thresher* sank in 8,500 feet of water), this was the practical limit for the bell. Link and his colleagues recommended that a manned vehicle be developed, possible with existing technology, that could do rescue work at 6,000 feet and could be flown anywhere in the world within twenty-four hours. Such a craft, called a DSRV, will be available in 1970.

The DSSRG group also had civilian uses in mind for this vessel. For one thing, the Navy during the more than fifty years it has had submarines has lost only two of them in peacetime. The crew for the DSRV would find time heavy on their hands unless they had other work to do. It was seen that the craft could be used for research, exploring, oil prospecting and drilling, mining, fishing, underwater cultivation, or for salvage.

Improved methods of underwater navigation were also recommended as well as research to find some practical way of raising the large bulk and weight of a submarine.

Ed Link, extending his thinking from the problem of lost subs, came to the conclusion that to do many desirable things at great depths man needed to know much more about living and working there. He thought of all the wealth that could be used once mankind learned how to survive and work in the sea.

After the group wrote its final report, Link flew to the *Sea Diver,* which he had arranged to have delivered to him in Key West. To replace the cylinder he had used in the Mediterranean, he had, between meetings in Washington, developed a "submersible, portable, inflatable dwelling," encased in rubber, which naturally earned the name of SPID. Robert Stenuit, the Belgian Diver, was skeptical when he saw the rubber casing. His doubts proved well founded. One day a glass bottle of milk, left forgotten by a workman inside the rubber housing, fermented and exploded. Flying glass pierced three holes in the wall. A new, sturdier dwelling was ordered, with walls reinforced by nylon.

Link's new experiments required a two-man team because one man could not keep track of all the instruments nor stay awake on watch for the many hours planned down in the deep. Jon Lindbergh, a young man with considerable diving experience, was hired as Stenuit's partner. Stenuit was on his guard before Lindbergh's arrival because the competence of the man alongside you is extremely important when you are deliberately getting into dangerous situations. The two men found that they worked exceedingly well together as a team, which was fortunate because they were in for a certain amount of trouble.

What was the new goal? In Link's words, "Call it the deepest long dive. Call it the longest deep dive. . . . To put two men a long way down for a long time—more than 400 feet for more than 48 hours—and thus to probe beyond the limits of human experience." The ultimate purpose? To learn how men can *live and work* at depths of 600, even 1,000 feet, and discover how to exploit the fabulous resources of the Continental Shelf.

Before the dive began, SPID, the dwelling, was lowered into the reasonably warm waters of the Bahamas and anchored on the bottom at a depth of 432 feet. Then on the morning of June 30, 1964, Stenuit and Lindbergh went overboard from the *Sea Diver* and swam into the cylinder that Link had designed and that would carry them down to their underwater dwelling. All seemed to be going well. They hit the bottom at one o'clock, gradually increased the pressure within the cylinder to that of the sea outside, opened the hatch, and swam out and 15 feet over, to SPID. Once inside, they went to work turning on all the instruments, which had been disconnected in case SPID flooded on the way down. The lights burned only five seconds and then went out. Suddenly, a sealed-beam light bulb exploded and sprayed the side of the housing with glass fragments. When Stenuit plugged in the radiator, it did not work.

By flashlight, Stenuit attempted to start the machine that would filter out carbon dioxide. When he turned the correct valve, nothing happened. There was a spare carbon dioxide filter in the housing, but someone had put on the wrong cover and so it had no valve at all.

Without any way to purify the air, without light, without heat, the two men nevertheless decided to stay down for at least twenty-four hours.

Stenuit signalled their predicament to the *Sea Diver* by Morse code. (The helium-talking problem remained unsolved.) Link signalled back that he was sending down a line to which they should attach the defective filter. It would be replaced on the surface and returned.

Stenuit and Lindbergh waited four hours in the dark for its return. When it came back, to their relief the valve worked. Through the chilly night they spelled each other on watch.

The next morning they began to swim outside SPID, breathing through fifty-foot lengths of hose attached to the housing. At such depths aqualungs are of no use because the tanks of gas would be consumed too fast. Before dinner time Lindbergh had managed to repair the radiator and they had fixed the lights. That night was more comfortable but frequently interrupted by a large fish called a grouper which kept bumping into SPID as he blundered after smaller fish which were attracted by the light shining through the porthole.

The following morning the two men went out again into the sea to do some exploring. Shortly after noon a message came from the surface. They had just passed forty-eight hours—two days and two nights—at this incredible depth. All the experiments had succeeded. Stenuit and Lindbergh would gladly have continued on, now that all their problems were solved, but they were ordered to surface. They swam over to their cylinder, locked themselves in, and gradually they were brought back to the surface.

Decompression in a chamber aboard the *Sea Diver* took four days. It would have taken no more time, however, if they had stayed down several months. The blood's nitrogen-saturation level had been reached. Decompression could be no worse for those who followed them, and learning from the troubles of Stenuit and Lindbergh, their successors would have a much easier time of it. Link and his brave divers had very substantially advanced the progress of man in the sea.

The "retired" inventor could not simply rest with this achievement. Along with the Union Carbide Corporation and the Binghampton Company that produces the Link Trainers, Link formed a new group called Ocean Systems, Inc., "To engage in commercial development of a broad range of underwater services and supporting systems." Robert Stenuit and Jon Lindbergh were soon on the payroll.

Link believed that putting men in the sea could be a paying business. Helping drill underwater oil wells would be profitable immediately. Other realistic applications for the present might be construction work in the sea, such as building bridge foundations and laying pipe. Some time in the future Ocean Systems might help in such undertakings as fish farming.

The company's first announcement said that it was "opening a new frontier that's every bit as challenging as that of outer space—but one that is much closer to man's grasp."

As Ocean Systems went into business, Link remarked, "This will be the first commercial operation organized to exploit with men in the sea the boundless wealth of our oceans. It is more than sustained deep diving, or diving within the confines of submarines or diving chambers. It is occupation of the oceans—not just in shallow waters but the deep oceans of the continental shelves, which add more than ten percent to the world territory and resources of our earth."

Link and his company soon found a variety of business. When the Air Force lost an H-bomb in the sea off Palomares, Spain, Ocean Systems was called upon to coordinate the work of eighteen civilian contractors hired to help find it. The company has also handled other deep-sea projects for the government: salvaging nine torpedoes from a depth of 3,000 feet, locating "an unarmed weapon" off the east coast of Puerto Rico, and finding a jet aircraft lost in a lake in Oklahoma. A Norwegian company has employed them to salvage the Nazi battle cruiser *Bluecher,* which was sunk when invading the Scandinavian country in 1940. Ocean Systems also made many dives to survey for the U.S. Navy the best site for Dr. George Bond's Sealab III near San Clemente, California. The company is now routinely performing engineering work below 400 feet and has forty divers trained to work at this depth.

Striving to go down and down farther into the sea, two of Link's men have now tested life in a pressure chamber set up to resemble conditions in 650 feet of water. They lived at this pressure for forty-eight hours and two minutes. During this experiment the men breathed a neon and oxygen mixture for thirty minutes with no apparent harm to their systems. "Each recited Lincoln's Gettysburg Address in an understandable voice, a marked improvement over the high-pitched, garbled speech in the helium-oxygen environment."

Recently Link has developed a two-chamber submersible called

*Deep Diver* which operates down to a depth of 1,350 feet. The man in one of the chambers remains substantially at sea level pressure, as in a submarine. The other chamber, with an atmosphere similar to that outside, has a diver exit, or lockout, and from it men can emerge into the sea. No diver is routinely doing this yet at depths much below 400 feet, but there is no doubt that they will. With Edwin Link in charge they will certainly continue to "probe beyond the limits of human experience."

# 7 Townsend Cromwell

## *Neptune's Child*

*Knowledge of wind- and earth-driven ocean currents first became important when men began to sail across the Atlantic. Spanish galleons took a southerly route to cross the sea to Central and South America to take advantage of nature's westward thrust in the latitudes just north of the Equator. Returning with treasure, they sailed north through the Florida Straits to catch the Gulf Stream bound for Europe. In the narrow passages between the Bahamas and the mainland of North America they were often caught by pirates and hurricanes. Currents might be useful but could also create hazards.*

*Sailors from the American colonies learned soon after landing at Plymouth Rock to take advantage of the Gulf Stream. It took British captains many, many years to learn why the passage to America took weeks longer than the return voyage.*

*Because there has been much less traffic in the Pacific the currents there are less well understood. In general, though, they obey the same laws of physics as those in the Atlantic. All major currents on the world's surface were identified one hundred years ago.*

TOWNSEND CROMWELL, a lean and bronzed young surfer-turned-oceanographer, had no idea when he sailed from Honolulu in August of 1952 that this voyage would put his name in the textbooks. To Cromwell the expedition in the Pacific was just one of many he had taken during the last three years—a bit more interesting, however, because the mission was scientific rather than commercial. Usually he earned his pay as oceanographer attached to a Federal Commission whose job was to improve tuna fishing in the sea.

Because the idea for the present voyage had been his, Cromwell bore the title of leader, but in practice he shared all important decisions with two scientific and personal friends who had come along.

These men were Professor R. B. Montgomery, on vacation from Brown University in Rhode Island, and E. D. Stroup, an oceanographer from Hawaii also employed to find tuna fish in the Pacific. The scientist friends, with a crew of fifteen, traveled aboard the *Hugh M. Smith,* a 700-foot tuna clipper that belonged to the U.S. Fish and Wildlife Service.

The object of the trip was to investigate reports that long fishing lines cast into the waters along the Equator were behaving unusually. A few months before, a research fishing ship had gone to the Equator to look for tuna. After stopping its engines the ship began to drift to the west, predictably carried by the Equatorial Current that flows from the Americas toward Asia. What surprised the fishermen was that the lines, 150 feet down in the water, did not drift along with the ship but moved rapidly toward the east, in the direction opposite that of the current.

Cromwell received permission to organize a scientific voyage to study this curious behavior and so came upon a brand new ocean current—a stream whose volume was that of a thousand Mississippis. No one teaching the story of the sea had ever heard of it before. After the discovery the three young scientists aboard the *Hugh M. Smith* agreed to name their finding the Equatorial Undercurrent. That the stream at last came to bear Townsend Cromwell's last name is due to sentiment. He was killed a few years later in an airplane crash near Acapulco, Mexico, while on his way to another oceanographic voyage.

In a more romantic era it would probably be written that Cromwell was born with the sea in his blood. No one can show such a thing scientifically, but it is certain that the earliest memories his family have of him always show him in some relation or other to the sea.

Townsend Cromwell was born in Boston in 1922. His older brother, Richard, has a snapshot of him toddling down a beach while he was learning to walk. His earliest letter to his mother (dictated to his father because he was still too young to write) tells her how he just caught his first fish.

While Towny, as everyone called him, was still a little boy, his family moved to La Jolla, California. This small beach town just north of San Diego is one of the loveliest in the United States. Excellent swimming exists almost the entire year round. It is also the home of the Scripps Institution of Oceanography.

Considering the influences around him, Towny's choice of a career hardly causes surprise. He turned constantly to the sea. While still in grade school, he once took a skiff out into the Pacific and stayed out there alone all day until darkness finally brought him home. His loving family, who had great confidence in him, worried not at all about his absence. Not the least bit talkative, Towny never told them whether he had gone out to get solace from the sea for some childhood distress or just to experience the joy of being on the water.

Years later he sat on a beach with his new wife, watching the waves break. "You know," he said a little sadly, "you can never enjoy it the way I do."

His brother Richard, seven years older, took up surfing and deep-sea fishing long before Towny could manage such sports. He wanted to follow of course, and did so as soon as he was big enough. Once he could handle the board and get through the big waves, he became one of the outstanding surfers on the West Coast. When he was old enough to hold a summer job he became a lifeguard on the beach at La Jolla.

During his sixteenth summer, though, he went to work in the woods, helping to clear the grounds around a Palomar mountain cabin. From there he wrote a letter to a friend of the family:

> Dear Mrs. Blankenship. I've read it in books, I've seen it in the movies, and I've heard people tell about it, but I didn't think there was anything to rival the ocean. This is it. Mrs. Kenyon's cabin is built in a sheltered valley midst oaks, firs, cedars, pines. They are easily one hundred feet high. Looking westward down the ravine: ridges and mountains.
>
> "From the way I've written, you might think the ocean has dropped to second in my estimation. No, the waves along the seashore still say more to me than the soft breezes that whisper through the trees. But in the same way the ocean sinks to slumber in the evening, so do the trees become silent and the air still at sunset.
>
> "I'm not the least bit homesick. The thing I miss most is the ocean. (Surfing.)"

For college Towny chose the University of California at Los Angeles. The beach at Santa Monica is only a few miles away. World War II interrupted his education, and he joined the Army Air Force, which trained him to become a meteorologist. Since the sea is the source of much of our weather, the military made a fortuitous choice.

Before going overseas Towny wrote to his family from Jefferson Barracks, near St. Louis. After describing his status, he said:

> "Excuse me if I reminisce for a minute. I have just listened to a program of dreamy Hawaiian music and my surfing days are vivid in my mind. What a wonderful, healthy, fine feeling old Savage and Forshaw and I used to have. I'm suppressing some desires within me—mainly to get out on a boat and work like a dog with no shirt on and to go relax on the beach and surf for weeks on end. I'm going to do that as soon as I get home."

Towny served in Great Britain and Germany, and at the end of the war he had become a captain. Back in the United States he did take the vacation he had planned, but not for as long as he thought he might have. Increasing maturity made him realize that life required more of him than just being a great surfer.

Returning to U.C.L.A. he received his bachelor's degree in June, 1947, then drove one hundred miles south to La Jolla to enter the Scripps Institute for graduate study in oceanography. On Christmas Day that year he married Katherine Huchthausen from Chicago, even though she did not know how to swim. She was nineteen years old. He was twenty-four.

His wife has a note Towny received on January 8, 1948, fourteen days after his wedding, from H. U. Sverdrup, at that time Director of Scripps Institute.

> "Memo to Messrs. Cromwell, etc.: I wish you to take part in a cruise during the next week, beginning Monday, January 12, and lasting probably five days. Your participation in the cruise cannot be taken as a regular part of instruction. However, you ought to get experience in work at sea."

Towny scribbled a note to his wife on the bottom of this order. "Darling—Read above and you will see why I have gone surfing. Be back two or three or four days." Before going out to work on the sea, the young student felt a need to caress it in his own familiar way. The new Mrs. Cromwell was finding out that she would always have to share her husband's love with an old mistress—the Pacific Ocean.

The following year Towny received his Master of Science degree from Scripps. Now he was a professional, the U. S. Department of Fisheries offered him a job as Oceanographer of the Pacific Ocean

Fishery Investigation in Honolulu, Hawaii. The new scientist flew out alone, leaving his wife and two infants in La Jolla for the time being. His letters to her reveal a lonely husband but one who was able to find some consolation. The first day he reported for work, his new boss suggested he might take the day off to get settled.

To his wife, he wrote:

> "I agreed and after settling myself in approximately 10 minutes, I went to Waikiki. The surf was only two or three feet high but it was a thrill to ride with the warm trade winds blowing about 20 knots off shore. You ride straight into it and you feel as if you're going very fast. Tomorrow I start work in earnest."

A two month's cruise at sea was being planned and Towny worked very hard. The ship for the cruise went out to test equipment. The oceanographer reported:

> "We saw a huge marlin swordfish leaping and playing about 100 yards from the boat. They're beautiful and shimmer in the sunlight when they jump. Also saw my first large tropical shark. It looked eerie, suspended in the deep blue water right under the bow."

Before the ship actually sailed, Towny bought a secondhand car for $75, which was about as much as the father of two could afford. One day he wrote:

> "A little native beach boy wanted to go out to Waile so I asked him to ride with me and show me the way. (The natives are really friendly but they are very cautious with strange *hoales*—people from the Mainland.)
>
> "We got to Waile and the surf was very big. About 10 feet and sort of vicious. Anyway, I went out and the surf was a howling screamer. When I went in, the boy cornered me and told me what had happened when I went out.
>
> "His story: 'When you go out I say he's good *hoale* surfer and all the beach boys they laugh and say "We'll see. We'll see." And then when you caught that wave—Woooo-eeee! They don't laugh no more.'
>
> "Actually I was very conservative all day and tried to keep out of everyone's way. Today when I went out again they were

EDWARD DIXON STROUP

*Townsend Cromwell in his favorite habitat*

all very friendly and talked to me a lot. It is much more fun surfing with friends."

Throughout his life Towny was always unusually well-liked. His wife says he was so special because "he had this power of self and of self-control." R. B. Montgomery, in the brief obituary he wrote about Cromwell for a scientific journal, explained that "his work was characterized throughout by close collaboration with his associates, who became his devoted friends. His personal charm and modesty, together with his scientific interest, led a number of oceanographers to join his field of activity."

In his last letter of this period, the husband wrote, "Darling. I will be glad when you are out here. You will like it fine, even the surfing part." This descendant of Neptune was trying gently to convert his non-aquatic wife. To arouse her interest in the new world where she would soon live he described Makaa Beach: "The water is zircon blue and the swells are green and the shallow water is clear and transparent. It's a beautiful beach to sit on or read on and the bay is a surfless paradise."

Then the *Hugh M. Smith* stood ready for sea, and Towny became happily active doing real work rather than talking about it or writing about it. When he returned from the voyage, devoted to finding undiscovered areas that might make good tuna fishing grounds, his wife and children had arrived in Honolulu. They rented a small house and for the moment the letters cease.

Cromwell continued to surf, and he went to sea four months out of every year. His wife understood all this and did not complain. Today she says, "Towny was something like a glacier. A subsurface man. Words had no meaning for him. Actions did. He would say, 'Don't tell me. Show me.' He wouldn't accept anything he couldn't measure. To him this indicated integrity, something he couldn't live without. And he never lost his temper."

With such an appreciation of her husband, it was perhaps easier for Katherine Cromwell to live the somewhat lonely life of the wife of a sailor.

The voyages of the *Hugh M. Smith* took Cromwell over a wide part of the Pacific and around an area as large as the United States. Wherever he went he kept his wife posted. He found letters very difficult to compose and thought little of himself as a writer. But consider this description of Oahu as he looked at it from the sea:

> "The beauty is indescribable. Palms line the shore and the slopes leading up to the mountains are dense jungle green. I've never seen the tops of the mountains. They are always hidden in the base of cumulus clouds which swell up thousands of feet into the sky. This evening at sundown they were soft pink. Tonight, under the full moon, they are silvery billows, as plainly visible as if it were mid-day."

On another occasion, at Canton Island more than a thousand miles south of Hawaii, he wrote:

> "This afternoon we are going pearl diving again. Yesterday we found five pearl oysters—but no pearls. The reefs where we are diving are the most beautiful I have ever seen. Every color and shade. There are purple mushrooms from the size of a silver dollar up to that of a dining room table. There are large coral

'trees'; green and blue and violet and multi-colored fish dart through the boughs. The water is warm and still.

"Tomorrow we are to put out for a week on long line fishing. This type of fishing has never been done here and since the water is infested with large tuna, shark, swordfish, etc., we should have a lively and interesting time."

In another letter, Cromwell tells about a trip to Hull Island.

"Our visit there was an event I'll never forget. Briefly—the island is beautiful, covered with palms, and the natives were trolling in outrigger canoes when we arrived. When our first skiff approached shore, the natives came running and laughing and screaming out of the palm-forest to meet them.

"In the midst of the great landing of the skipper, the engineer, and three others, a wave swamped the whole outfit and they washed up on the beach, full of water, amid the laughter of all the natives.

"Later, Herb and I and three others went in and the natives asked us to come to their meeting place. We sang 'You Are My Sunshine' and 'Home On The Range' and the natives all clapped. Then they sang some of their falsetto chants. Fred June, one of our biologists and quite a character, tried to show the chief's wife how to jitterbug. She did pretty well and everyone laughed until they almost split.

"We had a fine time. This island is almost never visited and is unspoiled by whites. The people are very friendly and courteous. They constantly brought us fresh coconuts to drink and eat."

In another letter from Canton Island he said:

"Underwater spearing is no good here—too many sharks. But spearing while wading is pretty good. I got two Uhus in about twenty minutes. One about eight pounds. Uhu is a very good eating fish but difficult to describe. Its color alone is unbelievable. It is bright green, blue, orange, red, and violet. We caught some sea turtles—turtle steaks are mighty fine."

While Cromwell's employer, the U.S. Government, remained interested in improving the nation's fishing industry, Cromwell grew more and more interested in how the physical conditions of the sea affect the fish population. The elements that make up the environment of the sea vary considerably from area to area. So do the kinds and quantities of fish.

As Cromwell and his colleagues roamed all over the ocean, they measured the saltiness of the sea, its other chemical elements, the temperature of the water at various levels, the plankton content (basic food source for all higher forms of marine life), and the various currents in the ocean. Their long lines with many hooks, which they dropped over the side at frequent intervals, helped them to take a census, particularly of the tuna population that in many parts of the sea had never before been studied.

After long, hard work aboard ship Cromwell would return to Honolulu for the even harder work of analyzing all the observations. His really pioneering labors greatly increased the sum of knowledge of the Central Pacific. As a result, a great deal more is known about these waters than about the Atlantic at comparable latitudes.

Having spent three years learning so much about the nature of his favorite ocean and being a scientist as well, Cromwell quickly recognized the possible importance of the news that at the Equator the Pacific seemed to break known rules. An untrained observer might have observed the long lines going in the wrong direction, hardly even wonder at it, and go on his way. Cromwell saw the report as an opportunity to do some basic investigating. Having talked his way into getting a month of the *Hugh M. Smith*'s time (even small ships cost upwards of $1,000 per day to operate) and having enlisted friends Montgomery and Stroup to go with him, he sailed south in August, 1952, to look into the matter.

When they reached the Equator at a point more or less due south from Hawaii, the oceanographers did not lower long lines for tuna but lowered drogues instead, which are cone-shaped, canvas sea anchors. These were weighted at the bottom and buoyed up at the top, so they would remain upright. Sunk to 150 feet below the surface, they were attached by lines to markers at sea level. These could be followed by radar. Another device, buoyed to remain on the surface to follow that current, was placed in the water at the same time. This also could be followed by radar. If the two kinds of surface markers moved significantly apart from each other, the current 150 feet below must be a different stream.

In a period of eighteen days Cromwell and his friends made twelve observations starting at 2° South of the Equator and continuing up to 7° North. As long as they stayed within 120 miles of the Equator, the surface marker moved to the west, as expected, but the drogue down at 150 feet always moved east at a speed of more than one knot. Farther north or south than this, the deep undercurrent could not be found. Where found along the Equator, the undercurrent proved to be long, fast, and remarkably thin—sometimes only a few feet thick from top to bottom—about 1,500 times as wide as it was thick. Sharply defined currents moving west flowed all around it.

How could such a current remain undiscovered one hundred years after every major surface current of the seas had supposedly been observed and described? (The new current later turned out to be almost as large as the Gulf Stream and thousands of miles greater in length.) The answer lies in the fact that all surface currents were discovered by drifting ships, and it was taken for granted that the water underneath behaved in the same way as water on top.

While steaming back towards Hawaii, the scientists began to speculate about the discovery. Why was this current so intimately connected with the Equator? To begin, they reviewed their general knowledge about currents. The well-known surface currents are formed partially by the prevailing winds. At latitudes near the Equator, these blow from the east. Hitting the continents of North America and Asia at mid-latitudes, the westward moving currents are turned north by the shape of the land and by the spin of the earth. This latter is known as the Coriolis Effect, the observation that water in the northern hemisphere spins to the right and in the southern hemisphere to the left. (This effects how water drains out of a sink. In the United States, for example, the water swirls in a clockwise direction, but in Australia the direction is counter-clockwise.)

At the Equator there is no Coriolis Effect. The spin of the earth exerts no force on the movement of water, and at a depth of 150 feet prevailing winds have no effect. These facts explain *how* the undercurrent could exist. They do not explain *why*.

Back at their various desks, Cromwell, Montgomery, and Stroup began to compose a joint paper on their findings. In the cautious way of scientists, publication in *Science* Magazine took two years. They wrote:

> "In part, the dynamics of the current seem clear. The surface layer of warm water deepens westward, so that throughout

> this layer there is an eastward component of pressure-gradient force. Below the depth of influence of the southeast trade winds, water flows east in response to the pressure gradient.* These considerations do not explain the narrowness of the current."

(In fact, though much more research has since been done on the phenomenon, all the questions raised by it are still not answered.)

Shortly after this voyage Cromwell was offered a position at the Scripps Institution, and he returned with his family to La Jolla. Because of the nature of his work, of course, he continued to make oceanographic cruises. The profession is not for stay-at-homes. While in Hawaii, in addition to four month cruises every year, Cromwell had had to go on a long trip to India and Ceylon to attend scientific meetings. There he saw so many fish and fish ponds that he wrote, "I don't mind saying I hate fish." Another time during the Hawaii period, he left for some months to study at Brown University with Montgomery.

From La Jolla he travelled south several times, pursuing a subject of great interest to him. This was the question, what causes great discontinuities of temperature in the ocean? That is, what brings about sharp boundaries between warm and cold water, and why do abrupt cold fronts exist?

He explained his interest in a report he gave in 1954 which was widely covered by West Coast newspapers. During a press conference held at Scripps, he talked about why fish concentrate in the frontal zones, or thermoclines, where cold and warm sea currents meet. He said, "Anything that is buoyant will not sink. It will just float over the convergence zone. The same thing is true for plankton [the microscopic bits of life that are the basis of the food chain]. . . . There are concentrations of plankton that are carried to the frontal zone from both water masses. Larger forms, including fish, follow the plankton to feed off them. We feel this is the reason why fish are found in large numbers near the fronts." He added that fishermen have long known about this and use surface indications, long lines of kelp and foam, as a guide to good fishing grounds.

Meanwhile, no one had forgotten about the undercurrent. In 1955 a two-ship expedition went out from Scripps to see if it con-

** Pressure-gradient force is caused because the large volume of water moving to the west must go somewhere at the end of its course. The path selected near the Equator is down and back toward the east.*

tinued on to the South American continent. Cromwell was in charge aboard the *Spencer F. Baird,* and John A. Knauss, a more senior oceanographer, had over-all charge aboard the *Horizon.* Cromwell took his ship to a station on the Equator at 115° West, due south from Lower California. Then he sailed on to Costa Rica, from where he wrote his wife:

> "The trip is going very well, I think, and the ocean has been obliging. It has done a number of things for us to measure. We found my current again flowing along the Equator, down deep, in the direction opposite to the surface current. We worked in the Equatorial Countercurrent for John Knauss—a dreary week of hard work with rather messy weather, but good results from both vessels and Johnny should be pleased with the results.
>
> "I've learned quite a bit on this cruise about expedition leading and I have considerably greater respect than I had before for the 'tough' approach. There are people, both scientific and on the ship, who won't work unless you make them. At first I was a little apprehensive but I'm happy to find I'm equal to the task and I believe that I could lead any kind of oceanographic expedition now. Darling, I'll write you next from the Europa Hotel, the ritziest in Costa Rica."

After a few days in port, the *Baird* sailed again and Towny found time for a letter:

> "Darling. This leg from Costa Rica to Peru has seemed very short after that first long stretch of thirty-eight days at sea from San Diego. We've had a lot of work to do but are now on our way in.
>
> "We were quite successful on this leg, too, I believe. We tested for the undercurrent and it was not there. It was not supposed to be."

(The theory had been that the undercurrent would be broken up by the mass of the Galapagos Islands and that there would not be enough ocean space between the Islands and the mainland, where these last stations were made, for the current to get in motion again.)

Once again at home he finished a paper he had been working on jointly with W. S. Wooster, a systematic analysis of all the data he had collected, titled *An Oceanographic Analysis of the Eastern Tropical Pacific*. Then, late in the year, he flew off to Thailand to attend a scientific meeting. He stopped off in Tokyo to hire a well-known Japanese oceanographer before returning to La Jolla for Christmas. All this activity makes it clear that Towny's star was rising fast in the scientific community.

In the spring of 1958 he went to sea again and wrote a note from Ascension Bay in Mexico (famous among biologists as a breeding place for whales):

> "We've had a keen trip, except for a few days. We've seen whales jump out of the water, sharks, caught thousands of fish, picked oysters off trees, and gotten burned. We should be in San Diego about 8:00 Friday morning, barring weather, breakdowns, etc. On the run home we are in the hands of the wind gods."

But such gods as there may be for travelers gave Townsend Cromwell little more time. His ship arrived on schedule, and then soon after his return, he prepared to go out again on another expedition.

While he was getting ready, a cable arrived from the Dolphin Expedition, led by John Knauss aboard the *Horizon,* stating that they had followed the undercurrent all the way from 140° West, along the Equator, to the western side of the Galapagos. The current had now been confirmed for a distance of 3,500 miles.

On June 2, 1958, Townsend Cromwell was aboard an airplane in Mexico en route to Acapulco with a companion, B. Shimada, to join another oceanographic mission. Near Guadalajara their plane crashed, killing both men. Cromwell's age was thirty-five.

It is not yet entirely certain that the Equatorial Undercurrent will always bear his name. He planned the cruise of the *Hugh M. Smith,* but other scientists aboard also had a hand in the observations. He never claimed the discovery as solely his own. And later research, in hindsight, showed hints that its existence was known before 1952. One observation of something like it in the Atlantic was made by an English scientist in 1886, and a Japanese scientist observed it in 1926—but these reports were published and forgotten. No one revived them until several years after Cromwell's voyage.

Whatever its name, research on the current continues. It was finally located in the Atlantic by ingenious use of a nuclear submarine in 1961 with characteristics almost exactly the same as in the Pacific. Further Pacific studies suggest it may run across the entire ocean at the Equator, a distance of more than 8,000 miles. Sought for in the Indian Ocean, it only appeared during the month of February.

Whatever becomes of his name in geography, an oceanographic ship called the *Townsend Cromwell* now exists. Built for the Bureau of Commercial Fisheries, it contains the most sophisticated, modern research gear available.

Mrs. Cromwell attended the christening ceremonies in Honolulu in 1964 with her five children. While there, a Honolulu reporter interviewed the family at the edge of a Waikiki hotel swimming pool and wrote:

> "Mrs. Cromwell hardly seems aware of the burdens she took on after the death of her husband in 1958. She has him to thank for it. 'Towny trained me very well,' she said yesterday. 'He had to be off and away on expeditions for two months at a time and he wanted us to get used to it. To him, the most important thing was for us to be very happy.'
>
> "She seemed very surprised at the suggestion that life had been rough on her. 'A friend of mine said after Towny died that we had had ten wonderful years together. I felt the same way.'
>
> "Jenie wanted to know if the ship would have been named for her father if he had not died. Her mother said 'No' and she frowned.
>
> "Dixon, her son, playing around the pool with his sisters, stopped for a moment near his mother. He said he liked to study the ocean, all right, but that social studies and English were his favorite subjects. He jumped in the water."

Townsend Cromwell's life was short but full and he had the satisfaction not given to every man of working in a field for which he had a passion.

# 8

## *Alvin*

*One of the United States' first small submersibles, designed to hover in the water at mid-ocean depths in order to observe the behavior of fish, suddenly found itself involved in serious problems of international relations. Like earlier American pioneers, the little craft acquitted itself with honor.*

NO ONE planned that *Alvin* would become a hero. The name in old High German means "noble friend" and that seemed to be destiny enough for *Alvin*. Actually, in this case, it was not really a true name at all but a contraction in honor of Allyn C. Vine, a resident genius at the Woods Hole Oceanographic Institution, who had a great deal to do with the inspiration and design of the small research submarine.

Allyn Vine, an Ohio youth who received his Master of Science degree from Lehigh University while Maurice Ewing taught there, went to Woods Hole in 1940 as a physical oceanographer. Some years later, fascinated by news of the Cromwell Undercurrent in the Pacific, he devised an ingenious and inexpensive means of checking its existence in the Atlantic. He induced the chief of Naval Operations to send a very strange order to the captain of a nuclear submarine that was then sailing home from the South Atlantic. The order called for the captain to position the submarine, when it reached the Equator, so that its bow would point up at a 45° angle and its stern would be 300 feet below the surface and then steer north for five minutes. But the puzzled captain found that once the submarine was in this position the vessel was forced to remain stationary—the pressures of the Cromwell Current and the current above it were exactly equal.

Dr. Vine became interested in submersibles again in 1964. This time he traveled aboard the French Navy's bathyscaphe, *Archimede,* when it descended four miles down into the deep Puerto Rican Trench. By visual contact he hoped to get some clue as to the cause of the Trench's existence. "It was like a kid bumping down a flight of stairs," he said. "We slid down the face of the Trench at a speed of only one knot. The slope was apparently just mud and ooze." Vine came up with no theory about the Trench but was impressed with the possibilities of personal observation. He felt, however, that bathyscaphes have severe limitations. The 70-foot yellow *Archimede* could only go slowly up and down over one spot, like an elevator. From it one could only get a bare minimum of the total ocean scene.

So *Alvin* came into being essentially because Dr. Vine and others like him had become excited about the possibility of using their eyes to study the ocean. The sea had always proved quite frustrating to scientists. From a ship on the surface, a man can only look down a few feet into the salty medium, but the most interesting things he wants to learn about may be going on thousands of feet below.

Much has been discovered about the sea from instruments and gadgets and hardware dropped to the bottom on long lines, but the samples are random, hit or miss at best, and important objects a mile below may be missed by only a foot or two. Electronic soundings merely give the average of the depths. They do not report small crevasses, hills or bumps that may be important obstacles when one is looking for things on the bottom.

If someone wants to know what things look like in the deep sea, why not simply lower a camera and take pictures? "Because of the prevailing turbidity of sea water," Allyn Vine pointed out, "photography has been a much more limited tool than sound in underwater work. The distances at which good photographs can be taken range from less than 1 meter (39 inches) in muddy harbors to perhaps 30 meters in clear water."

But, he went on to say, "The problems in underwater photography have not been optical so much as matters of handling and positioning—of keeping the equipment simple enough to handle over the side of a pitching platform and getting the camera and light source pointing at the right subject at the proper range. This is no small feat when everything is dangling out of sight on a current-swept cable 5,000 meters long.

"Unfortunately, the study of marine life at any depth by taking random photographs is as unproductive as a photographic study of birds would be if one took only random photographs of the sky."

This is one reason why some biological oceanographers are so interested in mobile, man-carrying submersible vehicles.

Here was part of the inspiration for *Alvin* and similar deep-diving research subs. No instrument can duplicate a human's ability to poke around and see for himself, so *Alvin* was planned to take scientists down as far as 6,000 feet in the ocean to do this human thing. No longer would cores, trawls, dredges, or cameras be the only means of taking samples. The deep sea could be studied all at once in its three-dimensional complexity. Scientists could selectively consider the particular rocks or sediments of marine animals that might help provide an answer for one particular problem. For all its small size, *Alvin* should permit ocean science to take giant steps forward.

With the blessing of, and about $750,000 from, the U.S. Navy, the little submarine was built by the Litton Industries in Minneapolis, Minnesota. Unveiling first took place at the factory there on May 21, 1964. A press release described *Alvin* as "a sophisticated 'tool' to further scientific underwater exploration, to be used in the study of biology, geology, and physical oceanography. The research crew will be able to observe marine populations at great depths, see more comprehensively the composition of the sea floor and collect samples of it, measure speed and direction of currents, as well as obtaining continuing profiles of temperature, salinity and other water characteristics."

Painted a bright banana yellow, *Alvin* is 22 feet long and weighs 13 tons. It looks something like a plastic toy duck designed for children to play with in the bathtub. It has a top speed of six to eight knots. Cruising speed in depth was expected to be two and a half knots but has turned out to be four knots. It has ten hours endurance and a range of 20 to 25 miles.

The pilot and his passenger have four portholes through which they can see ahead and beneath them. The pilot has a "joy stick," as in an airplane, by which he can operate the three propellers to make the raft go up or down, ahead or astern. An up-and-down-looking echo sounder measures the height of the *Alvin* above the ocean bottom and its depth below the surface. A scanning sonar set and a closed circuit TV system are used in addition to the viewing windows for obstacle avoidance. There are also a sonar telephone for communication with a surface ship, underwater cameras, strobe and incandescent lights, and, most usefully, a mechanical arm that can be extended out as far as 20 feet to do most, if not all, of the things a human hand can do.

WOODS HOLE OCEANOGRAPHIC INSTITUTION

Alvin, *world-famous small submersible, which can descend more than a mile into the sea*

*Alvin* arrived at Woods Hole on June 5, 1964. In a speech accepting it, Dr. Paul Fye, Director of the Institution, pointed out another research project for the vessel. "It may be used by ichthyologists for studies of the phenomena known as deep-scattering layers. These layers occur down to depths of several hundred fathoms. A peculiar characteristic of the layers is that they scatter, or reflect, sonar signals. They are generally seen on the recordings of echo sounding equipment and are thought to be caused by fish. Scientists have been unable to photograph these layers satisfactorily, and trawling leads to ambiguous results. *Alvin,* with her capacity to take a scientist into the middle of these layers, may prove to be the means whereby science can positively solve these riddles."

That summer the first careful shallow test-dives were made in waters off Cape Cod no more than 80 feet deep. Following extensive inspection and refitting during the winter and spring, *Alvin* went by land to Cape Kennedy, Florida, for more shallow water tests. In July, 1965, the vessel was taken to Coral Harbor on New Providence Island in the Bahamas. There it joined its mother ship, a catamaran, something that has been described as a cross between a houseboat and a dredge. A catamaran does not have much in the way of shape or lines to endear it, but *Alvin* urgently needs it for housing, to restore its batteries and to rest the crew.

The catamaran carried the small sub to the Tongue of the Ocean, a great canyon in the Bahamas between Andros and New Providence Islands. There it was lowered on the first very deep dive, without a crew, on a line that let it down to 2,340 meters (or 7,700 feet). When the submersible finally came up from a mile and a half below the surface, a round trip that took eight hours, an inspection showed it had weathered the test very well.

Then on July 20, William O. Rainnie Jr., the pilot, and Marvin J. McCamis, the copilot, went on board. Freed from any lines this time, *Alvin* started down, powered by the electric batteries. Now two men were risking their lives to find out if everything in the little ship worked properly at great depths and to see if they could get down and return to the surface without accident. When the sound gear showed that *Alvin* had reached 1,825 meters, the 6,000 feet originally planned, William Rainnie reversed the propellers, and the pilots returned safely to the surface.

These men and Valentine Wilson, a forty-one-year-old civilian from Falmouth, Massachusetts, concentrated during the trials on learning how *Alvin* performed and on how to operate a craft radically different from anything anyone had ever handled before. (Unlike a submarine, it has mechanical arms and a man can see out of it. Unlike a bathyscaphe, it can be maneuvered backwards or forwards.) Through their reinforced windows and with their powerful lights, however, they did find time to gaze into the deep and observe rugged cliffs and many exotic forms of plant and animal life.

After several weeks of testing in the Bahamas, the catamaran ferried *Alvin* to Bermuda, where it did its first actual oceanographic work inspecting the Navy's underwater installations near Argus Island.

In October the submarine came back to Woods Hole. The operation showed that *Alvin* could carry out the basic plans of the developers. The pilots came back safely—from depths that only a handful of men had ever seen before. The submarine was not perfect, however. Many problems in navigation, communications, battery life, and simple human comfort needed to be solved. With a revolutionary, new technology such things could be expected. Why else conduct tests? *Alvin* began to undergo some modification and overhaul.

Then, three months later on January 17, 1966, a U.S. bomber and a flying tanker collided during a refueling rendezvous at 30,500 feet in the sky while they were crossing over the Mediterranean coast of Spain. As the wrecks fell toward the earth, they passed through a 100-mile-an-hour jet stream which dispersed the various objects over a fairly wide area. As recounted in an official report later, "Aircraft debris, the bodies of seven airmen, and three or four unarmed weapons (H-bombs) carried by the B-52 bomber involved were recovered in and about the coastal village of Palomares. Impact with the ground caused low order conventional explosions in two of the weapons which damaged some buildings, but by good fortune there were no injuries among the population.

"One badly injured survivor from the B-52, who had been unable to free himself from his ejection suit, was found ashore. Three survivors, descending by parachute, were seen in the air and picked up by fishing boats from the water, the most seaward of them having been carried by the wind about nine miles bearing 095° true from the coast adjacent to Palomares.

"Considerable debris fell in the water as well as on land, and the master of a fishing craft, Francisco Simo-Orts, reported sighting a large parachute with a suspended object which landed near his boat and sank immediately. Orts fixed this position and was able to return to it on request. His description of what he had seen tallied with the hypothesis that this was a portion, or possibly all, of a bomb casing."

The accident involved Air Force planes and personnel, but the problem was handed to the Navy because by Act of Congress this service must "conduct salvage operations for all government activities." Several days' search suggested strongly that one U.S. H-bomb lay somewhere in the sea near the coast of Spain. The Navy went into action on January 22.

Almost three years before, when the submarine *Thresher* sank in the Atlantic, the Navy had discovered they had little knowledge or equipment to handle the situation. Experts were called in and recommendations were made. A long range program was started to remedy matters, but none of the new equipment was ready when the incident at Palomares caught the Navy by surprise. According to the Task Force report, "Few problems which arose had been previously experienced or evaluated, and all had to be solved under conditions of maximum urgency."

This meant, in effect, that almost anyone or any thing in the United States with working experience in deep water would be called upon. Included were the Reynolds International submersible, *Aluminaut,* which had just gone down to 6,250 feet and the 18-foot *Perry Cubmarine,* based on West Palm Beach, which was restricted to 600 feet. The day the Navy took over the search, a call also went out to

Otis Air Force Base near Woods Hole where *Alvin* lay, partially disabled, undergoing an overhaul. Technicians not only had to finish their work but then dismantle *Alvin* sufficiently so it would fit inside two Air Force planes. The conning tower had to be fitted into one cargo ship; the rest of the body, in another. In six days, nevertheless, *Alvin* was airborne and after delays due to bad flying weather, arrived in pieces at Rota, Spain, on February 5. The *Cubmarine* also flew across the Atlantic, but the 60-foot *Aluminaut* was too big and had to come overseas aboard an LSD (Landing Ship Dock) called *Plymouth Rock.*

Four days after arrival, reassembled, *Alvin* made its first Mediterranean dive, probing to 2,000 feet. Then it was loaded aboard the *Plymouth Rock,* too, to sail to the search area. At this point a mistral, a gale wind common to the area, blew up a storm with gusts of 70 miles an hour. *Alvin* suffered minor damage but was ready to dive on February 14.

*Alvin* was just one member of U.S. Navy Task Force 65 which finally grew to include 3,000 Navy personnel, 25 Navy ships, 4 research ships, 3 small submersibles, and many civilian contract personnel. The problems of the search were enormous. Few components in the huge force could do much work in the area of greatest interest—the deeply-fissured, rocky ocean floor, a half-mile deep, above which the Spanish fisherman had seen what might have been the bomb.

Before *Alvin* went to work, underwater demolition and explosive disposal teams had searched from the coast out to water eighty feet deep. They found much airplane debris but no bomb. The *Cubmarine* worked on the area down to 600 feet, without result. The search for the weapon by acoustical equipment turned out to be useless because the instruments were not sensitive enough to sort out the rugged details of the topography or distinguish a rock from the metal casing of a bomb.

When *Alvin* arrived on the scene it was assigned the area of highest probability, about ten to twelve miles out to sea. *Aluminaut* also had this area but left the deepest, roughest canyons to the smaller sub which could maneuver better in such restricted areas. *Alvin*'s pilots reasoned that the "object" could have slipped down quite a distance on the slopes of this terrain. Though their visual range was only 15 feet in the turbid water and the pitch black of eternal night, they searched along the bottom contours looking for the kind of track such a thing would have made as it scraped its way down a hill.

On the tenth dive, March 1, the pilots saw a furrow on a bottom slope that seemed to have the expected characteristics. They followed it down to 2,200 feet, then lost it because they had to return to the surface to replenish their batteries.

Not until March 13 was the track seen again, but *Alvin* had to surface once more because its batteries were exhausted. Troublesome as the batteries were, of course, problems of pin-point navigation in the absolute dark, with no reference points and little that anyone on the surface could do to help, also made this grim work. Trouble came, too, when the weather at the surface was bad and operations had to be halted.

On March 15, a month after their search had begun, McCamis and Wilson were in *Alvin* when the track was again located at a depth of more than 2,500 feet. They followed it down a seventy-degree slope, going backwards so that the line could be seen from the front windows without worrying if the propeller would snag on a rock behind them. Suddenly, there was a parachute wrapped around a cigar-shaped object of the right size, partially buried in about a foot of mud. A black and white photograph was taken, the news sent by sonar telephone to the surface, and *Aluminaut* ordered by the surface command ship to the scene.

Half a mile down, the two vessels talked easily with each other and rendezvoused one hundred feet apart. This was probably the first such meeting ever held in deep inner-space. *Aluminaut* had a pinger aboard which transponded to a signal sent from *Mizar,* the surface vessel, and now the weapon could be accurately located from above. *Aluminaut* could stay below for as long as twenty-four hours and did so, "baby-sitting" while *Alvin* surfaced to recharge. The "noble friend" emerged to find itself the object of world-wide acclaim.

But, once more, bad weather interfered. Another seven days elapsed before attempts to raise the object could begin. Now, however, a fix on its position had been obtained on the surface. *Aluminaut,* like *Alvin,* was supposed to have a mechanical arm but it was not fixed on at the time. When the storm let up, therefore, *Alvin* was sent down to take the new steps preliminary to lifting the 5,000-pound object.

The H-bomb was smooth surfaced. It had no hooks, rings, or any object on which a line could be conveniently tied. When *Alvin* went below, it carried an anchor and line from the *Mizar* which were to be used to grapple the parachute that was acting as a shroud. The mechanical arm managed to hook on to the parachute, but as the *Mizar* attempted to lift, the anchor cut the line. The bomb dropped and was lost for another nine agonizing days. Finally, after one more long intensive search in an area now riled by all the activity of small submersibles, strong currents and errant H-bombs, *Alvin* found the bomb again, 120 yards downhill and on the edge of a precipice that dropped into a crevasse at least 4,000 feet deep.

Then, after 135 dives on this project, the *Alvin* pilots got themselves into what might have been serious trouble. Overshooting the mark, the little sub got entangled in the parachute's lines. "It was a dangerous moment," Admiral Guest, in charge of the whole operation, recalled later. But Valentine Wilson, who was piloting, made a quick turn and *Alvin* was clear.

Some hours later a new object appeared on the scene. This was an unmanned vehicle named CURV, designed to recover lost torpedoes. It had been flown in from California. CURV is 6 feet high, 5 feet wide, and 15 feet long. It has a TV camera, three motors, lights, and recovery claws. It was suspended by a cable from a surface operating ship, in this case the U.S.S. *Petrel,* and was designed to operate at depths up to 2,000 feet. But the bomb was in much deeper water, so CURV could not be used until adjustments could be made for the greater depth.

CURV then went below in more or less the middle of the night, and *Alvin* guided it to its quarry. The unmanned vehicle was about to fix one of its large claws on the tube when it too became caught in the parachute. *Alvin* sent up word instantly and added that currents were sweeping both vehicle and bomb towards the nearby crevasse. Admiral Guest remarked to an aide, "I prefer combat to this any time."

But CURV had been able to hook on. The Admiral decided that even with a weak grip he had to begin to hoist. Slowly the bomb and CURV began to rise together, closely watched by *Alvin.* At 7:02 in the morning a new crisis appeared. The weight of the bomb and the waterlogged parachute were no longer evenly distributed on the lines, and there was danger that they would break, as they had before.

There seemed nothing to do, however, but continue hoisting.

At 8:19, only one hundred feet below the surface, the nylon lines of the parachute began to vibrate severely, threatening to collapse.

Navy scuba divers, already prepared, jumped over the side of the *Petrel,* swam down, and attached additional lines to the bomb. At 8:45 the weapon finally was hoisted on to the *Petrel*'s deck. Navy and Air Force men quickly disconnected its electrical system and defused its charge of TNT. Although the information is classified, it is certain that this bomb was at least fifty times more powerful than the one which destroyed Hiroshima.

Once the bomb lay on the deck of the *Petrel,* a very difficult public-relations decision had to be made. The sceptical world-press would have to be assured that the dreaded object really had been recovered. No photograph of a hydrogen weapon had ever been published, but how else than with a picture could the world be assured? At last selected members of the press were allowed on board, and pictures, taken from some distance away, were soon being radioed to newspapers everywhere. Pictures went out, too, of *Alvin,* unexpected actor in international affairs.

In a speech some time later, Secretary of the Navy Paul Nitze said, "The name *Alvin,* known only to a few in the oceanographic field, was flashed to the world, proclaiming it a hero at Palomares."

After *Alvin*'s return from Spain and the ordeal of the H-bomb, cracks and other signs of wear were noted in the little submarine. It had to be completely overhauled, delaying its scientific missions for some months. The opportunity was taken during this repair work to give *Alvin* new and improved equipment, including an automatic movie camera and a TV camera and light facing back toward the stern propeller so that the pilot could avoid obstacles behind him.

Finally, on June 14, 1967, *Alvin* arrived in the Bahama Islands aboard *Lulu,* its catamaran tender, to begin its scientific work. *Alvin* inspected bottom markers it had planted the year before and made a number of geology studies. Then it moved to a position off South Carolina where, on Dive No. 200, it confirmed the theory that the Gulf Stream reaches all the way to the bottom in that area. Woods Hole oceanographer R. M. Pratt, who was on that dive, reported, "Everything was in motion; sessile [bottom-fixed] organisms waved in the current, fish swam up into the current, and moving sand was causing the active formation of sand ripples, scour depressions and dunes."

On Dive No. 202 Alvin made contact with what the pilot thought was a six-foot "black rock." The story was published in *Naval Research Reviews* of November, 1967:

"The day began normally enough. It was Thursday, July 6, 1967. Skies were clear; the sea, calm. *Alvin,* the Navy's unique underwater research submarine, was launched from its catamaran and began its 202nd dive. Pilots Marvin McCamis and Valentine Wilson and Observer Edward Zarudski, a geologist from the Woods Hole Oceanographic Institution, made up the crew.

"Once on the bottom—a depth of 1,985 feet—the crew held *Alvin* in a stationary position while various scientific measurements were taken and recorded. McCamis had the conn; Wilson stood near the starboard window. Suddenly Wilson observed what looked like a six-foot-long black rock 30 or 40 feet away. When, a few minutes later, the pilots lifted *Alvin* off the bottom to an altitude of about five feet, the black "rock" came to life. As the dark shape charged the submarine at high speed, the crew could see that it was no rock but a swordfish, apparently angered at *Alvin* for disturbing its habitat. The fish closed in on the offending submarine, striking it in the joint between the forebody and afterbody immediately forward of the starboard window. Indeed, the force was so great that the fish's sword was driven in up to the hilt, about two and one half feet. As a result, the fish was securely wedged in the submarine. It was stunned for several minutes and then violently tried to withdraw.

"When the crew saw what had happened, they checked out all systems to determine whether any damage had occurred. The only abnormal indication was a slight intermittent leak detector indication. With the concurrence of Observer Zarudski and the surface controller, the pilots decided to terminate the dive immediately.

"When *Alvin* broke the surface, the swordfish was still securely trapped. However, in its efforts to free itself, the fish had nearly broken its sword through. Swimmers passed a line over its tail; thence to *Alvin*'s sail, and hoisted submarine, fish, and all aboard the catamaran.

"After a complete inspection and postdive checkout, it was determined that *Alvin* had not been damaged; there were only a few scratches in its Fiberglas skin. The sea denizen's attack had nothing to do with the slight detector indication.

"The swordfish was removed; succulent steaks were enjoyed by all. *Alvin,* shaken but unharmed, returned to the sea."

A major scientific role planned for *Alvin* was the investigation of the deep-scattering layer, a sonic phenomenon in the sea whereby a sound signal is returned to the sender as if it had hit a solid object in a spot where it is known no solid object exists. The theory had been that some kind of fish in schools caused this effect, but no one had been able to prove it.

On Dive No. 204 *Alvin* went down to look at a deep-scattering layer off Cape Hatteras. Here it found schools of squid, fish and a species of tiny crustaceans known as euphasids, each producing a reflecting layer.

Later, near *Alvin*'s home base at Woods Hole, the ship made Dive No. 226 to investigate a well-charted layer known to oceanographers as "Alexander's Acres." The surface echo sounder recorded the bottom of the layer to be 2050 feet down. *Alvin* descended to this depth and then, with its lights off but guided by its sonar, it approached the target. When the range was zero, the lights went on and the submarine found itself in the middle of a large school of thousands of lantern fish, each about a foot and a half long. *Alvin* had been rigged with a special net for this trip and it roamed through the school catching dozen of samples of this species which attracts its prey by a light which grows on its snout. *Alvin* proved beyond question what causes deep-scattering layers.

Anyone who has been on a ship, large or small, or even read about them, realizes that they all take on individual characters. In Navy terminology, *Alvin* does not deserve to be called a ship because it is so little that it can be carried aboard another vessel. Seen one way, it is a pile of metal, a collection of sophisticated instruments and machinery. Other small submersibles are being built along its lines, and the Navy is constructing a little nuclear-powered submarine called NR-1 which will be able to stay below for thirty days at a time. As of this writing, there are more than forty little submarines at work in various places in the world.

Why single out *Alvin* for praise? Because in many ways it was the first of its kind. It, and no other vessel, found the lost H-bomb, even though its primary mission in life was to aid science. In time there may be whole fleets of such craft, but like the *Santa Maria,* the *Mayflower,* the *Leviathan,* and the *Queen Mary,* the 13-ton *Alvin* is something special.

# 9 Columbus O'Donnell Iselin

## *The Captain Was a Scientist*

*If anyone could sail the waters of New England blind-folded, this would be the man. He has been able to indulge his passion for skippering boats and ships in the area ever since he was a boy.*

*At the same time he is a distinguished scientist, teacher, and for ten years headed a great sea laboratory. The two interests caused no conflict in his life. While still in college, he learned about the wonderful science of oceanography.*

THE FAMILY of Columbus O'Donnell Iselin, oceanographic pioneer, came originally from the village of Rosenfeld in the old German kingdom of Wurttenberg. From there, in the 15th century, the Iselins moved to Basel, Switzerland. Then in 1801, Isaac Iselin came to New York City and established a banking business which has flourished through a number of generations. The great-grandfather of the man who helped establish the Woods Hole Oceanographic Institution was himself one of the founders of the Metropolitan Opera House, The American Museum of Natural History, and the Metropolitan Museum of Art. The present oceanographer's first and middle names come from another ancestor, General Columbus O'Donnell.

Today's most prominent Iselin began his association with the sea almost from the date of his birth. Both his grandfather and father were bankers and yachtsmen who successfully defended the sailing "America's Cup" from the attempts of Sir Thomas Lipton to return it to Great Britain. When Iselin, a rich young man, entered Harvard University as a freshman in 1922, he was already a seasoned sailor.

This was the era of Gilded Youth, of Prohibition, and the Roaring '20s made famous in F. Scott Fitzgerald's stories and novels. With his background and opportunities, young Iselin might have become part of it. Instead he took up the serious study of mathematics, and for an extracurricular activity, the six-foot-three scholar rowed on the Harvard crew.

While in school he met Dr. Henry B. Bigelow, a zoologist and one of America's most prominent marine scientists. Dr. Bigelow channeled Iselin's love of the sea and his natural curiosity into a life's work. Within a few years the association with Bigelow led directly to the marine science establishment at Woods Hole.

But first, when Iselin received his bachelor's degree in 1926, he bought a yacht. By one account this was a confiscated rum-runner that he obtained from the Federal Government for $1,800 at a public auction. Another account is that Iselin had it built to order in Shelburne, Nova Scotia. Whichever is the case, the yacht measured 77 feet in length, weighed 37 tons, and was a sailing ship with gasoline auxiliary engines. Iselin named it the *Chance*.

He did not acquire the *Chance* as a plaything but for serious work. At the age of twenty-one he was going to captain his first oceanographic cruise. Hiring a professional cook, he invited eight Harvard classmates to sail with him as crew. Later, when the voyage was over, he had the schooner's log published as a book in a limited edition of 125 copies.

In it he listed the personnel:

Columbus Iselin II—21—Skipper.
Terence Keogh—20—Mate—A half-wit Irishman.
Robert Jordan—21—Engineer—Does not like Rolled Oats or Tomatoes.
John B. Shears—32—Cook—Deaf.
Oliver Ames—22—Crew—A Norwegian, strong as four horses.
John Churchill—22—Crew—That hard-boiled engineer.
John Knowlton—21—Crew—At least he is small.
Bert Hayes—23—Crew—Uses no clothes.
Robert Woodworth—23—Crew—The brains of the party, always busy.

The light-hearted crew sailed from New Rochelle, New York on July 3, 1926, after a sailing party attended by "Grandpa and Grandma Iselin and a good number of aunts." Iselin's mother and father followed for a few miles in another family yacht, and then the young men were on their own for a 5,000 mile voyage.

WOODS HOLE OCEANOGRAPHIC INSTITUTION

*Columbus O'Donnell Iselin standing in front of the Woods Hole Oceanographic Institution*

Labrador was their target. Professor Bigelow had loaned oceanographic instruments and shown the young sailors how to use them. The objective was to secure data on the upper Labrador Current and to study the local flora and fauna. On the second day out they were off New London. It was the Fourth of July and the boys celebrated by exploding numerous firecrackers, sometimes in each others' bunks. When a suspicious Coast Guard cutter approached, they let off more explosives to show that they were just a congenial bunch of fellows and not smuggling prohibited alcohol.

Sailing on, they crossed the Gulf of Maine (which Iselin was later to study in great detail). Off Halifax they tried to harpoon blackfish from the ship's dinghy but without success. They went ashore at the tiny French fishing island of St. Pierre, and Iselin, to his astonishment, was able to buy a whole case of brandy for the sum of four dollars. Continuing north, they found rough weather in the Cabot Strait and fog and icebergs on the Grand Banks off Newfoundland.

They landed again for gas and water at St. John's, which Iselin reported as "squalid, dead, and dirty." Nevertheless, they rented Room 412 in the New Foundland Hotel, and the entire crew took baths. That night the *Chance* once more set forth.

The first oceanographic stations, or stops to collect data, were taken in Whitefish Bay, Labrador. Then, continuing to head towards the North Pole, they ran into heavy ocean swells, some measuring 35 to 40 feet in height. The cabin, which could not be closed because the crew not on watch might suffocate, became very damp. Iselin, who was of course chief pilot, reported that he was spending a good deal of time dodging icebergs. "No wonder people do not go yachting much on this coast."

Now, when they landed, they encountered Eskimos. Ashore, some of the young men who were science students collected species of plant life to bring back to their professors. They also indulged in some mountain climbing, as well as trout and salmon fishing. At sea Iselin continued to trace the Labrador Current. At one point he lost his current meter overboard, but the engineers fashioned another one for him "to bring back to Dr. Bigelow."

Finally they reached Cape Chidley, Labrador, the entrance to the Hudson Strait. The young sailor had brought his friends safely to a point 60° North, the same latitude as the southern tip of Greenland. Then, on the first of September, they turned south again to sail home through these somewhat bleak seas. They stopped in Nova Scotia to wire all their various families that they were safe and well after a month of silence.

At last, on September 26, Iselin tied up the *Chance* at a dock in Boston. "There we were, after a 5,000-mile sail, and nobody particularly glad it was over. After three months of perfect companionship and a life without worries, the change to the big city is too sudden," he said.

But Iselin did return the tools to Dr. Bigelow and settled down in Cambridge to earn his Master's Degree in science. His findings on the course of the Labrador Current were used by the Coast Guard the following year to predict the course of icebergs.

During the three-month trip the crew had taken forty-five measurements of temperature and salinity at a depth of about 2,000 feet, netted many hauls of plankton, and collected specimens of plants and fish of previously unknown species. Iselin had also shown that the current moves an average of 12 miles a day south down the Continental Shelf along the whole coast of Labrador. Iselin was also pleased to prove that "serious investigations" could be done by a small boat handled by amateurs. "Previous oceanographic work," he wrote in his log, "has only been accomplished at great expense, both because of the use of large boats and because of professional crews."

In 1929 Iselin married and also became Harvard's associate curator of oceanography. During the summer he continued his Arctic expeditions. Of these, Thomas Barbour, Professor of Zoology, said, "I remember the joy and delight with which Henry Bigelow and I gloated over the first catches of beautifully preserved deep-sea fishes and invertebrates. From these days forward, Columbus has proved himself an inspired investigator and an excellent teacher, a painstaking and efficient executive and one whose research into the dynamics of the circulation of the waters of the ocean, and especially the Gulf Stream, have put him in the forefront of oceanographers of the day."

Dr. Bigelow had long felt the need for an oceanographic research center on the East Coast similar to the Scripps Institution on the Pacific. This was formed officially in January, 1930, at Woods Hole, Massachusetts, a small town on Cape Cod which was already a center for the study of marine biology operated by the U.S. Government. Funds for the new organization came from the Carnegie and Rockefeller Foundations. Dr. Bigelow became the first director of a center that has since become one of the largest of its kind in the world.

Columbus O'Donnell Iselin was informally associated with Woods Hole from the beginning, and his new yacht, *Atlantis,* served as its first ship. In 1931 a new research ship was ordered in Copenhagen for Woods Hole. Iselin was partially responsible for its design. When the new 142-foot, 420-ton ketch arrived at the Cape, Iselin sold his own *Atlantis* but retained the name for the new vessel. He was made captain of the special research ship and joined the Institution as physical oceanographer. The new *Atlantis* had two laboratories and "enough heavy cable to dredge the greatest depths and enough light cable to take temperature and salinity readings six miles down."

As Iselin once wrote about Woods Hole, "Until the Second World War the growth was gradual. Although our original building was always full in summer, there were seldom more than five or six investigators carrying on research in winter. The *Atlantis,* our fine new research vessel, cruised widely, especially in winter, and for the most part carried out three-dimensional surveys of the currents and of the chemistry of the water. Occasionally cruises were made whose primary objectives were the study of marine sediments, and a beginning was made in the development of geophysical techniques for exploring the underlying structure of the North Atlantic basin. The microbiology of the sea was developed to some extent, but few cruises were devoted to the collection of the larger deep sea forms.

"Nevertheless, by the end of ten years we had made considerable progress in understanding the currents of the western North Atlantic. In short, after the *Atlantis* had cruised for ten years we felt generally at home in most of the North Atlantic, both physically and chemically, and to a lesser extent, geologically and biologically."

Meanwhile, Iselin had begun teaching as a professor at Harvard, and in 1940 on Dr. Bigelow's retirement he became Director of the Woods Hole Oceanographic Institution. His particular research concern at the time was the Gulf Stream and the considerable variations in its behavior.

Then, "When the war clouds gathered, it was the many measurements of temperature and salinity, and their study, that chiefly paid off. With the early emphasis . . . on problems of undersea warfare, for the first time oceanographers found themselves consulted on matters of national defense by the scientific community of the nation as a whole.

"Woods Hole was visited by many scientists and engineers who previously had thought that this was purely a biological center. We taught a considerable number of Naval Officers the sorts of oceanography that play a part in both pro- and anti-submarine warfare.

"We acquired machine shops and electronic shops and gradually our full-time scientific staff came to include a hundred or so people, most of whom had previously had little or no connection with the sea. By the end of the war, our annual operating expenses had increased from roughly $150,000 per year (the approximate income from endowment) to about $1 million per year. The Federal Government had suddenly become a major source of support for our oceanographic laboratories."

Those who do not remember the impact of the bombing of Pearl Harbor personally may find it difficult to realize how worrisome the submarine menace seemed in 1941. Invasion scares frightened the residents of U.S. Pacific ports. Tankers hit by German torpedoes burned brightly just a few miles off Miami. From there north to Cape Cod the beaches were covered with the oil from sunken ships. At night Eastern coastal towns were dimmed out, and the beaches patrolled by sailors with dogs trained to attack suspicious characters. In spite of this guard, several Nazi saboteurs did manage to get ashore on Long Island from a submarine and begin to make their way toward New York City. Out in the Atlantic hundreds of ships with desperately needed cargoes for Great Britain were sunk by torpedoes made in Germany. For a time it looked as if submarines might decide the course of the war.

In the underwater emergency, the Navy of course turned to the men who knew most about this atmosphere. A good proportion of the experts in the field were connected with Woods Hole. There they labored for five years, perfecting existing devices and thinking up new ones, to help the United States win the submarine war.

Research centered on the astonishing character of sound in the sea, which varies according to the temperature, saltiness, depth, and the nature of the ocean bottom.

Working full time for the Government, men at the institution also pursued such problems as the underwater behavior of explosives, the development of anti-fouling paint for ships, and methods of locating life rafts. They worked out a means of forecasting what type of surf would hit any beach any place in the world at any time because knowledge of surf conditions is vital for amphibious invasions. Much of what the scientists did for the Navy is classified to this day, but the research and development of that wartime period so speeded up changes that any ship, surface as well as submarine, which survived the long conflict was obsolete by the time victory was achieved. For his wartime services, Dr. Iselin was awarded the United States Medal of Merit by President Truman.

Technically, World War II brought about an astonishing number of changes, and not only in oceanography. At the beginning few people knew about radar, jet planes, rockets, missiles, atomic energy, penicillin, snorkels, aqualungs, or jet streams in the upper atmosphere. These and many more such words were commonplace by the war's end.

The end of World War II did not bring about an end to international tension. It merely produced new potential enemies for the United States, and a new name, the Cold War. The Government realized that wars in the future would be fought as much in the laboratories as anywhere else. It began to invest heavily in scientific research. This decision made a considerable difference to the institution at Woods Hole. As Dr. Iselin said, "We now not only have research contracts with the Navy, with the Atomic Energy Commission, with the National Science Foundation, with the Weather Bureau, and with the Fish and Wildlife Service. There is worry that this form of support may not remain sufficiently stable to build up a strong scientific staff, but to date, with few exceptions, the policy in Washington has been remarkably liberal and understanding. Meanwhile, our own uncommitted funds have grown to about $300,000 per year. At the present time this is about 15 per cent of the total cost of our operations, but it is the key to their success."

The infusion of all this money has caused the Woods Hole Oceanographic Institution to grow in every direction. The town itself, especially in summer, has become too small. New scientific buildings have been built on a narrow neck of land near the ferries that take visitors to the resorts of Nantucket and Martha's Vineyard, and the resulting traffic jams are staggering. The ocean-going fleet of five ships now ranges familiarly all around the world, to study each one of the various specialties that make up marine science. Research projects now come in such variety as a study of the lightning caused by a new volcano in Iceland called Surtsey, a massive study of the Continental Shelf, a program to understand the life cycle of Atlantic salmon, and a search for the lost island of Atlantis in the Aegean Sea.

Dr. Iselin himself has branched out and lectures at the Massachusetts Institute of Technology, as well as at Harvard. After ten years he gave up his post as Director of Woods Hole, although he remains actively associated with the Institution. His thoughts continue to range widely over a number of vital areas.

About weather he has written, "During our lifetime we may be witnessing an example of one way in which human actions can affect weather and climate, through increasing the carbon dioxide content of the atmosphere.

"The atmosphere behaves much like the glass in a greenhouse. It easily transmits visible light but is rather opaque to the infrared or heat radiation coming from the ground and the sea surface. In the greenhouse the absorbing material is the glass roof. The corresponding material in the atmosphere is two substances present in minor amounts: water vapor and carbon dioxide.

"During the last 100 years, the increasing use of fossil fuels in our world-wide industrial civilization should result in the production of about 1,700 billion tons of carbon dioxide, 70 per cent of the amount now in the atmosphere. Because about two-thirds of the added carbon dioxide is absorbed by the sea, an increase of perhaps 20 per cent in atmospheric carbon dioxide can be expected.

"The effect of such an increase is not easy to predict, but there is reason to believe that it could result in a warming of the lower atmosphere by several degrees. Thus, we are conducting more or less in spite of ourselves a great experiment."

Commenting on the mysterious destruction of the atomic sub *Thresher,* he reasoned that a complicated series of meteorological events, caused by a great storm he had himself observed near the Gulf of Maine, created an exceptionally high internal wave, deep below the surface, which the diving officer may not have been able to handle. This would be something like a whirlpool which could carry the ship down at an alarming rate, before the officer knew what was happening. If any kind of electrical or mechanical failure occurred at this point, the sub would lose depth control and reach collapse depth before the trouble could be corrected.

On the future of oceanography he said, "It is the job of a laboratory like this and people like me to always be 20 years ahead, and I have been wondering how to start farming the ocean." He wants to know how to stir up the nutrient chemicals that are being continually drained downward, below the 100 meters at the surface where they could be utilized in photosynthesis. He foresees a threat to the sea's productivity in the disposal there of radioactive wastes. "We have already destroyed or used up much of the resources of the land. We should not, from ignorance or heedlessness, do the same thing to the sea."

Again about weather, he said, "Over what range of time does the ocean regulate our weather and climate. If we knew, we could make much more accurate long-range weather forecasts. Thanks to the large computers, this no longer seems a dream."

Concerning underwater acoustics Iselin said, "Sound is the one form (of energy) that travels well in water, and we should soon be able to make large areas of the ocean acoustically 'transparent.' By such means biological oceanography can be put on a quantitative basis. It will then no longer be necessary to guess how many fish were smart enough to escape the net."

About the future of certain forms of life in the sea he said, "Fish insure their survival by an enormous fecundity and it is usually economics rather than lack of science that automatically regulates the catch. With few exceptions, fishermen stop harvesting a given species long before the stock has been exhausted."

In an official biography Dr. Iselin reported that his areas of interest were "ocean circulation, underwater acoustics, and marine resources." These three realms overlap so much that they actually cover all the sciences of the sea. Thus, Iselin is a true oceanographer, since his concerns cover everything above, in and under the sea. At the same time, he has remained a dedicated sailor.

After some twenty years of service, it began to be clear that the *Atlantis,* which Iselin might forgivably call "my ship," was becoming old and obsolete. It had logged many hundreds of thousands of miles, and stood as platform for many discoveries. But the science had boomed so that it could not accommodate all the scientists who wanted to travel aboard nor was there room for all the new kinds of equipment. With several million dollars of Federal support, construction began on a vessel to replace the much-loved sailing ship owned by Woods Hole.*

* *In 1966,* Atlantis I *was sadly but proudly transferred to the Argentine Navy, to continue her work in the South Atlantic.*

When it reported for duty, the addition to the fleet was named appropriately, *Atlantis II*. At the time it was, at 2,300 tons, the largest ship belonging to any private research center in the U.S. Sentiment could not preserve sails for any modern working vessel. *Atlantis II*'s engines could cruise 8,000 miles at an average speed of twelve knots. Twenty-five scientists and a crew of twenty-eight could live in air-conditioned comfort in the tropics or dry and warm at the edge of the ice pack. It carried a glass bubble in its bow through which men could gaze out into the life of the sea. The ship contained four laboratories, hundreds of thousands of dollars worth of scientific equipment, easy-to-handle deck gear, and a pin-point navigation system. Her first major expedition, to the Indian Ocean, began on July 5, 1963.

After crossing to the Mediterranean and making a courtesy call at Monaco, the *Atlantis II* cruised through the Suez Canal to the Red Sea. There she made a station—to look into a phenomenon first discovered by her predecessor. This was the "hot spot" or "funny water," a very salty and lifeless area about a mile down, which has temperatures far above normal for any sea. It lies in pools several miles wide in the same latitude as the holy city of Mecca. The new ship confirmed its existence, then moved on to join in the International Indian Ocean Expedition. While in the "unknown ocean" or the "forlorn sea," as the Indian Ocean has been called, *Atlantis II* made 227 oceanographic stations. It returned home by way of South Africa and pulled into Woods Hole at one thirty on the morning of December 20, covered with ice.

When *Atlantis II* made its next, almost similar, trip, Dr. Iselin met it in Capetown for the long voyage home. As usual, being at sea gave him a sense of exaltation. This was the longest trip he had taken on the new ship, and he appreciated the comforts, the good spirits of the staff and crew, and the absolute perfect weather. They sailed on a long northwesterly-slant all the way to Bermuda with trade winds and fine weather, the wind dead astern all the way. Except for a visit to the island of St. Helena, the run was unbroken. Dr. Iselin's scientific purpose was to learn how to predict the occurrence and severity of diurnal warming (the daily heating of the sea's surface), but from his account, the sailing itself seemed to interest him just as much as his work. Columbus O'Donnell Iselin was once more combining two of his great pleasures in life—in the most ideal kind of way.

# 10 Taylor Aldrich Pryor

## *"... the nation will enter the sea."*

*The western frontier of the United States is reported to have closed before 1890. True, vast amounts of free lands were no longer available, but many other frontiers have since opened up: new businesses, new technologies, new sciences. Here is a man who is promoting the nation's newest science in its farthest western out-post and newest state. Further, he is working on the problems of man's communication with another species of mammal, a frontier certainly as challenging as the Rockies were to the pioneers.*

TAYLOR ALDRICH PRYOR fell under the blue-green spell of Hawaii while flying a helicopter for the Marines. "Tap," as he is almost always called, grew up in the not-very tropical state of Connecticut and went to college at Cornell University in upstate New York, where the winters can often be very cold. In 1953 he graduated and also married Karen Wylie, daughter of the author, Philip Wylie. During that busy year Tap also enlisted in the United States Marine Corps. He put in for the aviation branch and received his wings the following year. His life pattern was unconsciously set when his orders took him to the Islands out in the Pacific.

The very blond and dynamic young man saw not only the fabled beauty of the place but also a great opportunity that had barely been explored. Honorably discharged as a captain, he took up graduate study at the University of Hawaii in Honolulu. The school had recently begun to give courses in the relation of the ocean to science and these courses were what interested Tap. Karen Pryor, meanwhile, advanced in her chosen career of biology.

Although Tap's interest included everything to do with the sea, his researches primarily covered the behavior of sharks, whales, and porpoises. In 1959 he went to Eniwetok as a research assistant in a study of sharks that was sponsored by the Office of Naval Resea. and the Atomic Energy Commission. In 1959 he also began to pu. his dream into living reality.

He saw Hawaii, a chain of 122 volcanic islands completely surrounded by and right in the middle of the Pacific, as an unexcelled laboratory for exploration of the seas. The waters are warm and clear, marine life is abundant, and all desired depths quickly available. Eighteen thousand feet of water can be found less than twenty miles offshore. Tap Pryor felt that all these resources together spelled something very challenging. He also believed that the United States was progressing too slowly in marine research, that too much of it was compartmented, that a broad approach combining all the sciences was needed for important advances, and that the study of the sea would never be satisfactory until men got themselves into it. Further, he felt that the public was so poorly educated in oceanography that they did not give adequate support to conservation and research measures.

The year Hawaii became a state (1960), Tap was twenty-eight, and that year he established the non-profit Oceanic Foundation. He involved the new state government in his project by taking a long-term lease on 118 acres of state-owned land. His lease stated that the Oceanic Foundation must "operate a marine exhibit and research facility and devote a percentage of the gross income of the public exhibit to endow the research center." The Foundation must also be active in marine conservation and educate Hawaii's residents and visitors. It would seem as if Tap himself had written these conditions. And it was good to have local political backing for the status it gave the exhibit and the status it gave Tap as a fund raiser.

The land Tap chose at Makapuu Point was ideal for his purposes. He wanted a vast amount of unpolluted water, great depths nearby, room to expand, and all this near to the city and the University's research facilities. (Makapuu Point is a thirty minute drive from Honolulu.) In addition he wanted a spot of great beauty. Backed by green-covered cliffs and hills, a sparkling blue ocean in front, and several handsome islands offshore to frame the whole scene, Makapuu fulfills the role perfectly.

With the land secure, Tap now had to interest investors in his project. He considers himself a "concurrent scientist, administrator, businessman, politician, and community action director," a very useful combination of professions for an enthusiast trying to create something unique—a research institution supported by admission fees from the tourist public.

Since finance was the first objective, Tap began by raising more than a million dollars to build Sea Life Park. This somewhat resembles various Marinelands on the mainland, but it is much more science-oriented. Among the unusual exhibits is a 300,000 gallon, three-story, glass-enclosed coral lagoon, with a living coral reef and thousands of appropriate fish. Another is a million-gallon pool with an island and a half-size replica of the *Essex,* an old whaling ship of the type which used to call at Hawaii. This pool is also the home of the trained whales and porpoises. Karen Pryor, who supervises the training of these cetaceans, often narrates the show, a demonstration of porpoise-training techniques, the methods by which porpoises are released in the open sea and return voluntarily. Another attraction is an ancient Hawaiian fishing village rebuilt on its original site. A marine technology center, with closed circuit TV, shows underwater displays from offshore, working models that demonstrate various oceanographic principles, and a variety of undersea hardware. Opened in 1964, Sea Life Park now attracts some 300,000 visitors a year, and dividends are being paid to the investors.

Next on Tap's program was the Oceanic Institute, a research center for biology and ocean engineering. Almost the first project paid for by Sea Life Park was the education of Keiki, an infant porpoise. While little Keiki was being trained in a tank, someone swam with him for at least an hour every day. Soon the baby porpoise became very tame, even affectionate. When Keiki developed into a young adult, he was released from a boat in the open sea. He swam about until, hearing a familiar underwater sound signal, he returned to the Pryors in their boat. He had been made entirely dependent on the researchers for food and for affection. While in the sea, he entirely ignored all the numerous fish he could have eaten. The unaccustomed freedom probably frightened him as well, and so he responded promptly to the signal that offered safe and familiar surroundings.

Porpoises are highly social animals which, when free, live together in schools. As in other species of higher mammals, the young are dependent on adults for a considerable time. The Pryors wondered whether a school of porpoises would accept a stranger in their midst or attack and drive him away. The Pryors also questioned whether or not a tamed porpoise would learn to catch his own food before he

CAMERA HAWAII

*Tap Pryor, Hawaiian oceanographic enthusiast, with several of his trained porpoises*

starved in his new liberty. They did not try to find out with Keiki, but later they did release another trained porpoise in the midst of a free school of his species, and they were gratified to discover he was allowed to remain. If man can ever succeed in understanding the porpoise mode of talking, an animal like Keiki might become a messenger to his free fellows. This possibility, though it may seem remote, opens up great avenues of man-animal cooperation in fishing, exploration, and even, unhappily, undersea warfare. For this reason, Tap's Oceanic Institute was given a contract by the Naval Ordnance Test Station to support marine mammal research.* Specifically this includes studies of cetacean sonar, learning, communications, natural

* *At the Hudson Laboratories of Columbia University in New York studies are being made on how porpoises can aid in the recovery of sunken submarines, lost hydrogen bombs, and airplanes that have crashed in the sea. At Point Mugu, California, they have been trained to attach wires to sunken objects.*

history, social behavior, physiology, and applied uses of porpoises.

The largest sea farming, or fish culture, project in the world has also been instituted by Tap and his colleagues. This is going on at Makapuu and also on the island of Molokai, where ancient Hawaiian fishponds are being restored. The art by which the first settlers cultivated this source of high protein has been lost, but Tap hopes to rediscover it. If the yield of the sea in food can be substantially increased, and the knowledge exported to the nations now so perilously close to starvation, Tap Pryor's name should go down in history.

In addition to creating an immensely popular tourist attraction and Hawaii's first major research center, Tap has found time, among other things, to be elected a member of the Senate of the state of Hawaii. During the time the site for the Mohole project was being selected, Tap was very active in seeing that it be brought to his waters. The project to drill a hole through the ocean floor down to the earth's mantle and thus answer many riddles about the earth's past, did finally settle on Hawaii. Then it was canceled, due partially to the cost of the war in Vietnam but also due to the charge of "politics" in the selection of the contractor. Tap saw that the Mohole, with all the influx of scientists it would have brought to Honolulu, could have put Hawaii square in the center of the oceanographic map. The project's end disappointed him, but he hopes that in time it will be revived. Meanwhile, President Johnson has appointed him a member of the Commission on Marine Science, Engineering and Resources, and he will certainly bring attention there to the special facilities his state offers to a national oceanographic program. This appointment, which has caused him to open an office in Washington, D.C., and his various fund-raising trips to the mainland, plus his political duties, sometimes make him his own organization's bottleneck.

Nevertheless, Pryor's most ambitious project, the Makai Range, has gone ahead fairly well on schedule. In Hawaiian, Makai means "toward the sea" but what he is actually doing is going down into the sea. Like Dr. George Bond, Captain Jacques Yves Cousteau, and Edwin Link, he thinks the future lies under the surface of the sea. Unlike previous experiments, Pryor's installations will be permanent. Off Makapuu Point, he has built two underwater laboratories and plans more in the future. The first permanent living facility is at a depth of 70 feet, the second at 230 feet. Later there will be housing down to 600 feet and also a fully instrumented deep range, a "bugged" piece of the ocean from the shore down to 18,000 feet. A brochure describing all this offers "an onshore-offshore facility available to all industry and institutions for development and testing of equipment, instruments, and techniques at all depths," as well as "low cost access to the sea."

The list of programs and projects for which the Makai Range, its sea floor laboratories, and ocean testing facilities are being developed is an impressive one. Earlier deep-sea living projects centered greatly on the problem of survival. Pryor believes that the problem of keeping alive has now essentially been solved. The time has come to move on, for men to work routinely and productively in the ocean. Among the labors he suggests, those involving the oil industry would probably be carried out first.

"Future oil-well groups will be located offshore throughout Continental Shelf depths. Wellheads, piping, and monitoring equipment will be placed on the sea floor by divers using techniques of saturation diving; during construction, divers will live on location in submerged habitats for periods of thirty days. Control information will be telemetered to the wellheads by direct wire or submerged transponder buoys, and monitoring information from the wells will be transmitted to a central surface location. Remote control devices on the wellhead will control oil flow direction, quantity, and pressure. Oil-carrying pipes will lead to storage tanks. Divers will return to the well sites for maintenance, repair, and additional construction." Pryor's project "will provide onshore-offshore testing facilities for the development of these new tools and techniques and for the training of industry divers."

In addition to the oil industry, the Makai Range would be useful for studies in mining, undersea construction, more studies in improving underwater living, rescue and recovery techniques, and communications. For science, it could be employed for more studies on the human factors involved in such exotic living and for testing all the factors involved in humans exploring in 18,000 feet of water.

The installation at 70 feet has somewhat the shape of a piano. The other, at 230 feet, is smaller and looks like the safety valve on a pressure cooker. Both provide "reasonably comfortable" quarters for crews of four to six men. They breathe rare gases, have electricity, fresh water, and communications and data links through "umbilicals" from shore. The two housings are reached by "submersible work decompression chambers." When the crew's work is done, the men are brought to shore in this vehicle to a larger chamber for the detoxification period. Tap is working on a process by which he hopes to reduce the time it takes to bring a diver's blood back to normal. Ultimately he hopes to develop underwater habitats that are completely self-supporting. They will use nuclear fuels, cells or storage batteries for power and obtain their fresh water and breathing gases from the sea itself.

Other goals include the conservation of Hawaii's marine resources through zoning inshore reef areas to prevent pollution and over-fishing, the development of undersea parks, and "dramatizing the new era of man in the sea by sending a small, fast manned submersible from Hawaii non-stop along the ocean bottom to San Diego."

"But our real goal is to create an industrial and scientific critical-mass. There is a demanding opportunity today for a consortium of operating companies. . . . Somebody can cause companies to create these consortiums and exploit the sea. The institute's facility . . . will be indispensable. If as a private institution in these vacuum areas in the ocean spectrum we stimulate others to compete with us, we're achieving our goal."

Already his activities have stirred others to expand their efforts in the area. Tap has talked to the officials at Woods Hole about setting up an annex in Hawaii, and the project is now being studied. He has also discussed joint research projects with the University of Wisconsin and the University of Hawaii. The latter's Department of Oceanography quadrupled in size between 1964 and 1967. Its research projects include compilation of the Atlas of Physical Oceanography of the Indian Ocean, studies of pre-historic changes in the Pacific's sea level, the patterns of ocean currents around Hawaii, and sea-air interaction in the North Pacific. Many businesses are attracted by all this activity, and at least a dozen new ones have been established in Hawaii since the Sea Life Park opened its doors in 1964.

According to Pryor, "We envision Honolulu as an ocean-oriented scientific and industrial community with components mutually supporting and mutually stimulating. We envison Hawaii as providing a focal point through which the nation will enter the sea."

And there is Makapuu in the vanguard, "probably the first major oceanic research and development installation which practices the belief that oceanography is not a single academic science but a whole world."

# Suggested Reading

Briggs, Peter, *Water: The Vital Essence.* New York, Harper & Row, Publishers, 1967.

Bruun, Anton, ed., *The Galathea Deep Sea Expedition.* New York, The Macmillan Company, 1956.

Carson, Rachel, *The Sea Around Us,* rev. ed. New York, Signet Science Library, 1961.

Coker, R. E., *This Great and Wide Sea.* New York, Harper & Row, Publishers, 1962.

Gaskell, T. F., *Under the Deep Oceans.* New York, W. W. Norton & Company, Inc., 1960.

Hill, Maurice N., ed., *The Sea,* Vols. I, II, III. New York, John Wiley & Sons, Inc., 1963.

Raitt, Helen, *Exploring the Deep Pacific.* New York, W. W. Norton & Company, Inc., 1964.

Spilhaus, Athelstan, *Turn to the Sea.* Racine, Wisconsin, Whitman Publishing Company, 1960.

Stenuit, Robert, *The Deepest Days.* New York, G. P. Putnam's Sons, 1965.

Stewart, Harris B., *Deep Challenge.* Princeton, New Jersey, D. Van Nostrand Company, Inc., 1966.

Sverdrup, Johnson and Fleming, *The Oceans.* New York, Prentice-Hall, Inc., 1945.

United States Naval Institute, *Ocean Sciences.* Annapolis, Maryland, 1964.

For information about available courses in oceanography, the following catalog is recommended: *University Curricula in the Marine Sciences,* published by the Interagency Committee on Oceanography, U.S. Navy Department Annex, Washington, D.C.

# Index


Air Force, U. S., 98-99, 100, 103
Alexander's Acres, 105
*Aluminaut* (miniature research submarine), 45, 99, 100, 101
*Alvin* (miniature research submarine), 45, 94-105
Ames, Oliver, 107
Anchovies, 18
Angler fishes, 54-55
Aqualung, 37, 72, 77
*Archerfish,* U.S.S., 34
*Archimede* (bathyscaphe), 94-95
Argus Island, Bermuda, 39, 98
Ascension Bay, Mexico, 92
Astronomical clocks, 30
Atlantic Ocean, 66
*Atlantide* expedition, 49-50
*Atlantis* (research vessel), 27, 60, 62, 65, 111, 115
*Atlantis II* (research vessel), 116
Atmospheres, pressurized normal and synthetic, experiments in, 36-45
Atolls, coral, 17, 19
Atomic Energy Commission, 113, 118
Atomic tests, 13, 28, 64

Bahama Islands, 103
*Baird, Spencer F.* (research vessel), 14, 19, 91
Ballast Point, California, 43
Balbour, Thomas, 110
Basalt, 61
Bathyscaphes, 94-95
Bathythermograph, 24, 26-27
*Berkone,* U.S.S., 40, 41
Bermuda, 65, 98
Bigelow, Henry B., 107, 109, 110-11
Bikini atoll, 19
Bille, Steen, 50
Biscay, Bay of, 7
*Bluecher* (Nazi battle cruiser), 78
Bond, George F., 34-45, 72, 78, 121
Brown, Mike, 63
Bruun, Anton, 46-57
*Bruun, Anton* (research vessel), 46, 57
Buoyant ascent technique, 38

Cables, submarine, 31, 66
Caesarea, Palestine, 71
California, University of, 12
Camera for deep-sea photography, 61, 67
Canton Island, 86, 87
Cape Kennedy, Florida, 97
Cariaco Trench, 23
Carlsberg Foundation, 47, 49
Carlsberg Ridge, 47-48
*Carnegie* (yacht), 11
Carnegie Foundation, 110
Carpenter, Scott, 40, 41, 44
Cephalopods, deep-sea, 52
*Challenger,* H.M.S., 7, 9
*Chance* (yacht), 107-10
Churchill, John, 107
Coastal waters, 31
Cobb Seamount, 45
Cold War, 113
Columbia University, 58, 61, 62, 120*n*
Columbus, Christopher, 70
*Conrad,* U.S.S., 62
Conshelf programs, 37-38, 42, 43
Continental Drift, 17, 66
Continental Shelf, 31, 44, 59, 66, 110, 113, 122
Copenhagen, University of, 47
Cores and coring, deep-sea, 11, 64-65, 66, 68
Coriolis Effect, 89
Cousteau, Jacques Yves, 10, 34, 37-38, 39, 41, 43, 72, 121
Cromwell, Katherine Huchthausen, 83, 86, 93
Cromwell, Richard, 81, 82
Cromwell, Townsend, 17, 80-93
*Cromwell, Townsend* (oceanographic vessel), 93
Cromwell (Equatorial) Undercurrent, 13, 17-18, 81, 89-91, 92-93, 94
Crossroads, Operation, 13
Crust of the earth, 61, 62, 64, 65
*Cubmarine* (submersible craft), 99, 100
Currents, ocean, 32, 81, 88, 89-91, 92-93, 102, 109, 110, 111
CURV (unmanned vehicle), 102

*Dana* Expedition, 47, 48, 51, 54
*Daphne,* U.S.S., 56
*Deadelus* (frigate), 56
Decompression, 35, 37, 42, 71, 72, 73, 74, 77, 123
*Deep Diver* (two-chamber submersible vessel), 78-79
Deep diving, 76-77, 78
Deep-scatter layer, 105
Deep Submergence Systems Group (DSSRG), 74-75
Defense Department, U. S., 21
Department of Fisheries, U. S., 83, 93

Depth finders, sonic, 9, 14, 15, 67, 96, 105
Diver's osteitis, 42
Diving bell, 75
*Diving Saucer* (miniature research submarine), 38
Dolphin Expedition, 92
Donn, William, 67
Drilling the earth's crust, 65, 121-22
DSRV (manned vehicle), 75

Earthquakes, 13, 14, 16, 20, 62-63, 64, 67, 69; man-made, 61, 63
Earthquake waves, 64
Echo depth sounders, 14, 15, 67, 96, 105
Edmund Scientific Company, 28
Eels, 47
Eisenhower, Dwight D., 46
Electric sparkers, 61
Engineering, ocean, 32
Eniwetok atoll, 19, 117
Equatorial Current, 81
Equatorial Undercurrent, *see* Cromwell (Equatorial) Undercurrent
Ewing, John, 63
Ewing, Maurice, 58-68, 94

Falco, Albert, 10
Fiji Islands, 16
Fish and Wildlife Service, U. S., 81, 113
Fish farming, 78; *see also* Sea farming
Fishing industry, 13, 18, 26, 32, 80-81
Fitzgerald, F. Scott, 106
*Flip* (research vessel), 18
*Flying Fishes of the Atlantic* (Bruun), 48
Food problem, world's, sea as answer to, 9, 21
Fraser, F. C., 49, 50, 57
Fye, Paul, 97

Galapagos Islands, 91, 92
*Galathea* Expedition, 50-56
Galatheahauma, 54-55
Geophysics, 13
Globigerina ooze, 7, 19, 23
Great Acapulco Trench, 20
Guest, Admiral, 102
Gulf Stream, 89, 103, 110, 111
Guyots, 13

Hawaii, 117, 118, 121, 123
Hawaii, University of, 117, 123
Hayes, Bert, 107
Heezen, Bruce, 66, 67
Helium, 37, 39, 42, 72, 73
Herzen, Richard, 17
Holothurian, 53
*Horizon* (research vessel), 14, 15, 91, 92
Housing, undersea, 31, 38, 72-73, 75-77, 121, 122, 123
Hudson Canyon, 44, 66
Hudson Laboratories, Columbia University, 120*n*
Hull Island, 87
Humboldt Current, 32
Hydrogen bombs, recovery of, 78, 99-103, 105, 120*n*
Hydrophones, 61, 62

Ice Ages, 67
Indian Ocean, 46, 47, 51, 52, 56, 57, 93, 116, 123
International Council of Scientific Unions, 21
International Geophysical Year (1954), 28
International Indian Ocean Expedition (1963-1964), 9, 56, 116
International Oceanographic Commission, UN, 56
Iron ore, 23
Isaacs, John, 18
Iselin, Columbus O'Donnell, 7-8, 106-16
Iselin, Isaac, 106

Jamaica, 69
Jarl, Viggo, 49
Johnson, Lyndon B., 121
Johnson, Samuel, 16
Jordan, Robert, 107

Kennedy, John F., 28
Keogh, Terence, 107
Knauss, John A., 18, 91, 92
Knowlton, John, 107

Laboratories, underwater, 121, 122
Labrador, 109, 110
Labrador Current, 109, 110
La Jolla, California, 12, 34, 40, 81, 82, 90
Lamont, Thomas, 58
Lamont, Mrs. Thomas, 62
Lamont Geological Observatory, 17, 58, 61, 62, 63, 64, 65, 66, 68
Landslides, ocean, 63, 66
*Life Histories of Deep Sea Eels* (Bruun), 48
Life in the lower depths, 52-55
Lindbergh, Jon, 76-77
Link, Clayton, 72

Link, Edwin A., 69-79
Link, Marian, 69, 70
Link Flight Trainer and Simulator, 70, 77
Lipton, Sir Thomas, 106
Liquid breathing, 45
Litton Industries, 96
Loch Ness monster, 46
*Lulu* (catamaran tender), 97, 103

Magnetic poles, 68
Magnetism, 17, 68
Magnetometer, 17, 67
Makai Range project, 121-23
Makapuu Point, Hawaiian Islands, 118, 121, 122, 123
Mammal research, marine, 119-121
Manganese nodules, 13, 23, 33
Man in the Sea program, 39-45
Mantle of the earth, 17, 61, 65
Marianas Trench, 15
Marine biology, 12, 21
Marine resources, conservation of, 123
Massachusetts Institute of Technology, 25-26
Mazzone, Walter, 36
McCamis, Marvin J., 98, 101, 104
Mediagraph, 27
Menard, H. W., 17
*Meteorological Instruments and Weathercraft* (Spilhaus), 30
Meteorology, 26
Micropaleontology, 63
Mid-Ocean Ridge, 8, 63, 65-67
Minerals in the sea, 22-23, 32-33
Mining, deep-sea, 32-33
Minnesota University Institute of Technology, 24, 28, 33
*Mizar,* U.S.S., 101, 102
Mohole project, 65, 121
Mohorovicic, Andrija, 65
Molokai Island, 121
Monaco, 72, 74, 116
Montgomery, R. B., 81, 85, 88, 89, 90
Morgan, J. P., 58
Mountain ranges, undersea, *see* Mid-Ocean Ridge; Ridges
Murray, Earl, 40

National Research Council, 21
National Science Foundation, 57
National Sea Grant College and Program Act (1966), 33*n*
Naval Oceanographic Office, U. S., 74
*Naval Research Review,* 103-04
Navy, U. S., 9, 14, 24, 34, 35, 37, 38, 40, 62, 64, 72, 74, 75, 78, 96, 99, 100, 103, 105, 112, 113
Neopilina Galatheae, 53
Neutral buoyancy, 44
New London, Connecticut, 35, 39
New Providence Island, Bahamas, 97
New York University, 27, 28
Nitrogen, 37
Nitrogen narcosis, 35, 37, 42
Nitze, Paul, 103
North Pole, 18
NR-1 (U.S. nuclear-powered experimental submarine), 105

Oahu, Hawaiian Islands, 86
Oceanic Foundation, 118-24
Oceanic Institute, 119-20, 123
*Oceanographer* (Coast and Geodetic Survey vessel), 59
Oceanographers, 8-9, 16, 57
*Oceanographic Analysis of the Eastern Tropical Pacific* (Cromwell and Wooster), 92
Oceanography, 8-9, 13, 20, 24, 47, 58, 114, 123
Ocean Systems, Inc., 77-78
O'Donnell, General Columbus, 106
Office of Naval Research, 118
Ogdensburg, New Jersey, 63
Oil industry, 10, 78, 122
Oil wells, underwater, 78
"Our New Age," syndicated newspaper science feature, 25, 30

Paleoclimatology, 63
Palomares, Spain, 78, 99, 103
*Petrel,* U.S.S., 102
Philippine Trench, 51, 52-53
Phosphate, 32
Phosphorite, 23
Photography, underwater, 61, 67, 95
Pickford, Grace, 52
Plankton, 8, 41, 88, 90
*Plymouth Rock* (LSD), U.S.S., 100
Point Mugu, California, 120*n*
Population explosion, world's, 21, 23
Porpoises, 117, 119-21
Port Royal, Jamaica, 69-70, 71
Pratt, R. M., 103
Pressure-gradient force, 90
Protozoa, 19
Pryor, Karen Wylie, 117, 119
Pryor, Taylor Aldrich, 117-23
Puerto Rican Trench, 95

Radioactive heat, 19-20
Radioactive wastes, 13
Radioactivity in the sea, 13
Rainnie, William O., Jr., 98
Raitt, Helen, 14-15
Raitt, Ross, 14

Raw materials, sea as source of, 9
Red tide, 8
Rescue work, deep-sea, 38
Revelle, Roger, 11-23, 67
Rickenbacker, Edward Vernon, 62
Ridges, undersea, 8, 17, 20, 47-48, 63, 65-67
Rift valley, undersea, 67
Ritter, W. E., 12
Rockefeller Foundation, 21, 110

San Clemente, California, 78
Sardines, 13, 18
Sargasso Sea, 47
*Satellite of the Sun* (Spilhaus), 30
Schmidt, Johannes, 47, 48
*Science* magazine, 89
*Scientific American,* 22
Scripps, Miss E. B., 12
Scripps, E. W., 12
Scripps Institution of Oceanography, 11-12, 14, 17, 18, 20, 21, 40, 56, 67, 81, 83, 90, 110
Scuba diving, 35, 103
*Seacat,* U.S.S., 38
*Sea Diver* (yacht), 69, 71, 72, 73, 74, 75, 76, 77
Sea farming, 32, 114-15, 121
Sea food chain, 32
Sea-grant colleges, 31-33
Sealab programs, 39-44, 78
Sea Life Park, Hawaiian Islands, 119, 123
Seamounts, 16, 17, 19, 45, 66
Sea of Galilee, 71
Sea serpents, 56
Sea water, age of, 63
Sediments, 13, 19, 60, 64-65, 66, 67
Sediment temperature probes, 68
Seismic shooting, 60-62
Seismic waves, 63
Seismographs, 59-60, 63, 64, 68
Seismology, 14
Seychelles Islands, 52
Shears, John B., 107
Shellback Expedition, 14, 16
Shimada, B., 92
Simo-Orts, Francisco, 99
*Smith, Hugh M.* (tuna clipper), 81, 86, 88, 92
Sound physics, 62
Sound research, underwater, 18, 26, 61, 112, 115
South Seas, 16
Space clock, 25, 28-30
Speech modifier device, 39
SPID (submersible, portable, inflatable dwelling), 75-77
Spilhaus, Athelstan, 24-33
Stenuit, Robert, 73-74, 75, 76-77
Stroup, E. D., 81, 88, 89
Submarines, escape from, 38
Sulphides, 23
Surtsey (volcano), 113
Sverdrup, H. U., 83

Tecktites, 68
Temperature in the ocean, 90
Tharp, Marie, 66-67
Thermal gradient, 31-32
Thermoclines, 26, 27
Thompson, C. Wyville, 7
*Thresher,* U.S.S., 38, 74-75, 114
Tidal waves, 13
Tonga Islands, 14, 17
Tonga Trench, 14, 15-16
Tongue of the Ocean, Bahamas, 97
Topography of the ocean floor, 20
*Toro,* U.S.S., 38
Transocean bottom expedition, 31
Trenches, 14, 15-16, 20, 23, 51, 52-53, 95
Truman, Harry S, 113
Tuckfield, Chief Engineman, 38, 40
*Turn to the Sea* (Spilhaus), 30

Undersea break-away, 41

Vaquier, Victor, 17
*Vema* (research vessel), 61, 62, 63
Villefranche Bay, Monaco, 72, 73
Vine, Allyn C., 94-95
*Vityaz* (research vessel), 57
Volcanoes, underwater, 16, 19, 20, 67, 113

Water problem, world's, 22
Weather, study of, 10, 114, 115
Weather Bureau, U. S., 113
West Pakistan, 22
Whales, 50, 117, 119
Wilkie, Charles, 63
*Williamsburg* (yacht), 46
Wilson, Valentine, 98, 101, 102, 104
Wolff, Torben, 50
Woods Hole Oceanographic Institution, 9, 26, 27, 61, 94, 97, 98, 106, 107, 110-16, 123
Woodworth, Robert, 107
Wooster, W. S., 91
*Workbook on Meteorology* (Spilhaus), 30
Workman, Robert, 35
World War II, 112-13
Wylie, Philip, 117

Zarudski, Edward, 104

**DATE DUE**

| | | | |
|---|---|---|---|
| Oct 25, 1991 | | | |
| | | | |
| | | | |
| | | | |
| | | | |
| | | | |
| | | | |
| | | | |
| | | | |
| | | | |
| | | | |